Lecture Notes in Mathematics

Edited by A. Dold and B. Eckmann

789

James E. Humphreys

Arithmetic Groups

Springer-Verlag
Berlin Heidelberg New York 1980

Author

James E. Humphreys
Department of Mathematics & Statistics
GRC Tower
University of Massachusetts
Amherst, MA 01003
USA

AMS Subject Classifications (1980): 10D07, 20G25, 20G30, 20G35, 20H05, 22E40

ISBN 3-540-09972-7 Springer-Verlag Berlin Heidelberg New York
ISBN 0-387-09972-7 Springer-Verlag New York Heidelberg Berlin

Library of Congress Cataloging in Publication Data. Humphreys, James E. Arithmetic groups. (Lecture notes in mathematics ; 789) Bibliography: p. Includes index. 1. Linear algebraic groups. 2. Lie groups. I. Title. II. Series: Lecture notes in mathematics (Berlin) ; 789. QA3.L28. no. 789. [QA171]. 510s [512'.2] 80-12922

Printed in Germany

Printing and binding: Beltz Offsetdruck, Hemsbach/Bergstr.
2141/3140-543210

PREFACE

An arithmetic group is (approximately) a discrete subgroup of a Lie group defined by arithmetic properties - for example, $\mathbb{Z}$ in $\mathbb{R}$, $GL(n,\mathbb{Z})$ in $GL(n,\mathbb{R})$, $SL(n,\mathbb{Z})$ in $SL(n,\mathbb{R})$. Such groups arise in a wide variety of contexts: modular functions, Fourier analysis, integral equivalence of quadratic forms, locally symmetric spaces, etc. In these notes I have attempted to develop in an elementary way several of the underlying themes, illustrated by specific groups such as those just mentioned. While no special knowledge of Lie groups or algebraic groups is needed to appreciate these particular examples, I have emphasized methods which carry over to a more general setting. None of the theorems presented here is new. But by adopting an elementary approach I hope to make the literature (notably Borel [5] and Matsumoto [1]) appear somewhat less formidable.

Chapters I - III formulate some familiar number theory in the setting of locally compact abelian groups and discrete subgroups (following Cassels [1], cf. Weil [2] and Goldstein [1]). Here the relevant groups are the additive group and the multiplicative group, taken over local and global fields - or over the ring of adeles of a global field. One basic theme is the construction of a good fundamental domain for a discrete group inside a locally compact group, e.g., $\mathbb{Z}$ in $\mathbb{R}$, or the ring of integers O_K of a number field K inside $\mathbb{R}^n$ (n the degree of K over $\mathbb{Q}$), where a fundamental domain corresponds to a parallelotope determined by an integral basis of K over $\mathbb{Q}$. In the framework of adeles or ideles such fundamental domains have nice arithmetic interpretations. Another basic theme is strong approximation. These introductory chapters are not intended to be a first course in number theory, so the proofs of a few well known theorems are just sketched.

Chapters IV and V deal with general linear and special linear groups, emphasizing "reduction theory" in the spirit of Borel [5]. Here one encounters approximations to fundamental domains (called "Siegel sets") for $GL(n,\mathbb{Z})$ in $GL(n,\mathbb{R})$ and deduces, for example, the finite presentability of $GL(n,\mathbb{Z})$ or $SL(n,\mathbb{Z})$. The BN-pair (Tits system) and Iwasawa decomposition are used heavily here. There is also a brief introduction to adelic and p-adic groups.

Finally, Chapter VI recounts (in the special case of $SL(n,\mathbb{Z})$) the approach of Matsumoto [1] to the Congruence Subgroup Problem, via central extensions and " Steinberg symbols". Here adeles and strong approximation play a key role, along with the Bruhat decomposition already treated in IV. Matsumoto's group-theoretic arguments, done in detail, lead ultimately to the deep arithmetic results of Moore [1], which can only be summarized here. (It is only fair to point out that $SL(n,\mathbb{Z})$ can be handled in a more self-contained way, cf. Bass, Lazard, Serre [1], Mennicke [1], and unpublished lectures of Steinberg. Special linear and symplectic groups over other rings of integers can also be handled more directly, cf. Bass, Milnor, Serre [1]. My objective has been to indicate the most general setting in which the Congruence Subgroup Problem has so far been investigated; in this generality it has not been completely solved.)

The various chapters can be read almost independently, if the reader is willing to follow up a few references. I have tried to make the notation locally (if not always globally) consistent. Standard symbols such as $\mathbb{Z}$, $\mathbb{Q}$, $\mathbb{R}$, $\mathbb{C}$ are used, along with $\mathbb{R}^{>0}$ (resp. $\mathbb{R}^{\geq 0}$) for the set of positive (resp. nonnegative) reals. If K is a field, K^* denotes its multiplicative group.

Chapters I - V are a revision of notes published some years ago by the Courant Institute. Chapter VI is based partly on a course I gave at the University of Massachusetts; class notes written up by the students were of great help to me. I am grateful to the National Science Foundation for research support, and to Peg Bombardier for her help in typing the manuscript.

J.E. Humphreys

CONTENTS

I. LOCALLY COMPACT GROUPS AND FIELDS

Here we shall review briefly the approach to local fields based on the use of Haar measure, together with the construction of the adele ring of a number field. For a full treatment of these matters, the reader can consult Chapter I of Weil [2].

§1. Haar measure

For the standard results mentioned below, see Halmos [1, Chapter XI] or Bourbaki [3, Chapter 7].

1.1 Existence and uniqueness

Let G be a locally compact topological group (each element of G has a compact neighborhood, or equivalently, the identity element e does). G acts on itself by left and right translations

$$\lambda_x : g \longmapsto xg$$
$$\rho_x : g \longmapsto gx^{-1}$$

(here the inverse insures that we get a homomorphism $x \longmapsto \rho_x$ of G into the group of homeomorphisms of the space G).

Define a <u>left</u> (resp. <u>right</u>) <u>Haar measure</u> μ on G to be a nonzero Borel measure invariant under all left (resp. right) translations. This means, by definition, that μ is nonzero, all Borel sets are measurable, $\mu(C) < \infty$ for C compact, and $\mu(\lambda_x M) = \mu(M)$ (resp. $\mu(\rho_x M) = \mu(M)$) for $x \in G$, $M \subset G$ measurable.

<u>Remarks and examples.</u>

(1) If μ is a left Haar measure on G, $\hat{\mu}$ is a right Haar measure, where $\hat{\mu}(X) = \mu(X^{-1})$ for all measurable X^{-1}. (Check!)

(2) If μ is a left Haar measure on G, $c \in \mathbb{R}^{>0}$, then $c\mu$ is again such (so Haar measure, if it exists, cannot be absolutely unique).

(3) If G is abelian, left Haar measure = right Haar measure.

(4) $G = \mathbb{R}$ or $\mathbb{C}$ (additive group): Lebesgue measure is a Haar measure.

(5) On finite products, the product measure is again a Haar measure (for example, on $\mathbb{R}^n$ or $\mathbb{C}^n$).

(6) On the multiplicative group $\mathbb{R}^{>0}$, dx/x is a Haar measure. (Verify this by integrating functions of compact support.)

(7) G may have essentially distinct left and right Haar measures. (See Halmos, p. 256, for the standard example.)

THEOREM. Let G be locally compact. Then G has a left (hence also a right) Haar measure, and (up to a positive multiple) only one.

Haar proved the existence part for G having a countable basis of open sets (1933). Later von Neumann proved uniqueness for compact G. The general case was completed by Weil and von Neumann.

Exercise. G locally compact, μ (left) Haar measure.

(a) G is discrete iff $\mu(\{e\}) > 0$.

(b) G is compact iff $\mu(G) < \infty$.

When G is compact, one frequently (but not always) normalizes μ so that $\mu(G) = 1$.

1.2 Module of an automorphism

The uniqueness part of Theorem 1.1 is more useful than may appear at first. Let G be locally compact, with left Haar measure μ. Aut G denotes the group of automorphisms of G (as topological group).

If $\phi \in \text{Aut } G$, and $X \subset G$ with $\phi(X)$ measurable, set $\nu(X) = \mu(\phi(X))$. Since ϕ preserves Borel sets and compact sets, it is very easy to see that ν is again a left Haar measure on G. By uniqueness, $\nu = (\text{mod}_G\, \phi)\mu$, where $\text{mod}_G\, \phi \in \mathbb{R}^{>0}$ (and this number is independent of the original choice of μ, again by uniqueness). Call $\text{mod}_G\, \phi$ the (left) module of ϕ.

Example. Let $\phi = \mathrm{Int}\ x : g \mapsto xgx^{-1}$ $(x \in G)$. Here write $\mathrm{mod}_G\ \phi = \Delta_G(x)$, so $\Delta_G : G \to \mathbb{R}^{>0}$ is a function, called the module of G. If $\Delta_G = 1$, call G unimodular (this means that left Haar measure on G is also right Haar measure).

Exercises and examples.

(a) We could also have defined a right module of ϕ. Prove that this equals $\mathrm{mod}_G\ \phi$.

(b) $\mathrm{mod}_G\ \phi \cdot \mathrm{mod}_G\ \psi = \mathrm{mod}_G(\phi \circ \psi)$.

(c) An abelian group is unimodular.

(d) Any automorphism of a discrete or compact group has module 1, so such groups are unimodular.

(e) Any semisimple or nilpotent Lie group is unimodular.

Besides the example $\phi = \mathrm{Int}\ x$, another sort of automorphism and its module will arise in §2 when we discuss locally compact fields.

1.3 Homogeneous spaces

THEOREM. Let G be locally compact, H a closed subgroup of G. Then there exists a G-invariant nonzero Borel measure on the homogeneous space G/H iff the function Δ_G, restricted to H, equals Δ_H; in this case, such a measure is unique up to a positive multiple.

When G is abelian, or G is semisimple and H discrete, etc., the hypothesis will be fulfilled. It is cases like these that will occupy us later.

§2. Local and global fields

Here and in subsequent sections we are following the approach of Cassels [1] (cf. also Weil [2, Part I], Lang [1, Chapter VII], Goldstein [1, Part 1]).

2.1 Classification theorem

By <u>global field</u> we mean either a number field (finite extension of $\mathbb{Q}$) or a function field (finite extension of $\mathbb{F}_q(t)$, t transcendental).

By <u>local field</u> we mean the completion of a global field with respect to an archimedean or discrete (always rank 1 for our purposes) nonarchimedean valuation. $\mathbb{Q}$ has the completions $\mathbb{R}$ and $\mathbb{Q}_p$ (for primes p in $\mathbb{Z}$); $\mathbb{F}_q(t)$ has completions (all nonarchimedean) isomorphic to the field $\mathbb{F}_q((t))$ of formal power series. To get all local fields we just take all finite extensions of the fields just named. (Finite <u>separable</u> extensions will actually suffice.) Some authors do not regard $\mathbb{R}$ or $\mathbb{C}$ as local fields. Also, some authors allow more general coefficients for function fields. However, our definitions are the appropriate ones in the present context, as the following well known theorem shows.

<u>THEOREM</u>. <u>Let</u> K <u>be a (non-discrete) locally compact topological field. Then</u> K <u>is a local field, in the above sense.</u>

<u>Outline of proof</u>.

(1) If $\alpha \in K^*$, multiplication by α is obviously an automorphism of the (additive) locally compact group K, so its module (see 1.2) is defined. We denote it $\mathrm{mod}_K(\alpha)$ and set $\mathrm{mod}_K(0) = 0$. This function $\mathrm{mod}_K : K \to \mathbb{R}^{\geq 0}$ is our candidate for a valuation on K.

(2) It must be seen that mod_K actually is a valuation (since square of absolute value occurs for $\mathbb{C}$, one must define "valuation" appropriately: cf. Cassels, §1). The multiplicative property is obvious. To see whether mod_K is archimedean or not one looks at the prime field ($\mathbb{Q}$ or $\mathbb{F}_p$) and studies the various possibilities.

(3) Local compactness of K implies completeness in the metric topology defined by mod_K; in particular, K contains a copy of the appropriate completion of its prime field ($\mathbb{R}$, $\mathbb{Q}_p$, $\mathbb{F}_p((t))$). Local compactness also forces K to be finite dimensional over this subfield, which finishes the proof.

Exercise. It will be seen shortly that local fields are indeed locally compact. Compute mod_K explicitly, e.g., for

$\mathbb{R}$: usual absolute value

$\mathbb{C}$: square of usual absolute value

$\mathbb{Q}_p$: $p^{-ord_p(\alpha)}$ (see Appendix).

This singles out for each local field a normalized valuation, which we will always use.

Exercise. K a local field, mod_K as above, μ = Haar measure on K (additive group). Then $\frac{1}{mod_K}\mu$ defines a Haar measure on K^*. (Cf. 1.1, example (6)).

2.2 Structure of local fields

Let K_v be a local field, as defined above, with valuation $| \ |_v$. We assume the reader is familiar with the basic algebraic properties of K_v; since the topological structure of $\mathbb{R}$, $\mathbb{C}$ is sufficiently known, we require v to be nonarchimedean. So K_v is a completion of a number field or function field at a "finite place"; for some facts about the former case (of main interest to us) see the Appendix below.

$O_v = \{\alpha \in K_v \mid |\alpha|_v \leq 1\}$ is called the ring of local integers, and is a principal ideal domain (PID). It has a unique maximal ideal $P_v = \{\alpha \in K_v \mid |\alpha|_v < 1\}$, which is generated by an element π_v of the underlying global field K with maximum value < 1. The residue field $k_v = O_v/P_v$ is well known to be finite. (For $K_v = \mathbb{Q}_p$, these objects are $\mathbb{Z}_p$, $p\,\mathbb{Z}_p$, $\pm$ p, $\mathbb{F}_p$.) In the following theorem we list those topological properties of local fields which will be of importance to us.

THEOREM. Let K_v be a local field, v nonarchimedean. Then O_v is an open (hence also closed) subgroup of the (additive) group K_v; O_v is the unique maximal compact subring of K_v; and K_v is a (non-discrete) locally compact field.

Proof sketch.

(1) The neighborhood $|\alpha|_v < \varepsilon$ $(\varepsilon > 0)$ of 0 in K_v always contains a sufficiently large power of the "prime" π_v, so the topology is non-discrete.

(2) P_v is obviously an open subgroup of K_v; since k_v is finite, P_v has finite index in O_v, which is therefore a finite union of open cosets.

(3) That O_v is compact follows from the fact that it is closed, along with the general principle (exercise): A subset of K_v is relatively compact (i.e., has compact closure) iff it is bounded relative to $|\ |_v$.

(4) Any subring of K_v containing an element α with $|\alpha|_v > 1$ (i.e., α not in O_v) contains all powers of α and hence is not bounded. In particular, O_v is the unique maximal compact subring of K_v.

(5) O_v is a compact neighborhood of the identity in K_v, so K_v is locally compact.

Since $\mathbb{R}$ and $\mathbb{C}$ are well known to be locally compact, we obtain the converse of Theorem 2.1: All local fields are (non-discrete) locally compact fields.

Appendix: Review of number fields and completions

(a) Besides the usual (archimedean) absolute value $|\alpha|_\infty = |\alpha|$, $\mathbb{Q}$ has a p-adic valuation for each prime p: If $\alpha \in \mathbb{Q}$, write

$$\alpha = p^{\mathrm{ord}_p(\alpha)} \frac{\beta}{\gamma}$$

where β, γ are integers relatively prime to p, and $\mathrm{ord}_p(\alpha) \in \mathbb{Z}$. Define

$$|\alpha|_p = (\frac{1}{p})^{\mathrm{ord}_p(\alpha)} \quad , \quad |0|_p = 0 \quad .$$

(We could replace $1/p$ by anything strictly between 0 and 1, without changing the metric topology, but Haar measure yields this normalized choice: see 2.1.)

It is obvious that:

$$\mathbb{Z} = \{\alpha \in \mathbb{Q} \mid |\alpha|_p \leq 1 \text{ for all primes } p\}$$

and

$$\alpha \in \mathbb{Q}^* \text{ implies } |\alpha|_p \leq 1 \text{ for almost all } p,$$

hence $|\alpha|_p = 1$ for almost all p.

(b) Let K be a number field, of degree n over $\mathbb{Q}$. There are n distinct embeddings of K into $\mathbb{C}$, r of them <u>real</u> (with image in $\mathbb{R}$) and the remaining $2s$ <u>imaginary</u>, s pairs of complex conjugate embeddings. (In the literature, the usual notation is r_1, r_2 rather than r,s.) Denote these $\sigma_1,\ldots,\sigma_r$ and $\tau_1, \overline{\tau}_1, \ldots, \tau_s, \overline{\tau}_s$, respectively. Combining each σ_i with ordinary absolute value yields an archimedean valuation on K, extending $|\ |_\infty$; each pair τ_i, $\overline{\tau}_i$, combined with the square of ordinary absolute value on $\mathbb{C}$, yields another such. The resulting $r+s$ valuations exhaust the extensions of $|\ |_\infty$ to K, and may be called <u>infinite</u> valuations.

On the other hand, each p-adic valuation $|\ |_p$ on $\mathbb{Q}$ extends in at least one (and at most n) ways to K; the resulting <u>finite</u> valuations $|\ |_v$ correspond precisely to the prime ideals dividing (p) in the ring O_K of elements of K integral over $\mathbb{Z}$, the <u>ring of integers of</u> K.

These exhaust the archimedean and discrete nonarchimedean valuations of K. It is well known that $O_K = \bigcap_{v \text{ finite}} O_v$ (defined below) $= \{\alpha \in K \mid |\alpha|_v \leq 1$ for all finite $v\}$. Moreover, if $x \in K^*$, $|\alpha|_v \leq 1$ for almost all v (hence $|\alpha|_v = 1$ for almost all v).

(c) K (as above) has an integral basis over $\mathbb{Q}$, i.e., a basis $\{\omega_1, \ldots, \omega_n\}$ consisting of elements of O_K, such that $O_K = \mathbb{Z}\omega_1 + \ldots + \mathbb{Z}\omega_n$. ($O_K$ is a free $\mathbb{Z}$-module of rank n, as are all fractional ideals $\neq 0$ in K.) With notation as above, the <u>discriminant</u> D_K of K is defined to be the square of

$$\det \begin{pmatrix} \sigma_1\omega_1 & \cdots & \sigma_r\omega & \tau_1\omega_1 \cdots \tau_s\omega_1 & \overline{\tau}_1\omega_1 & \cdots & \overline{\tau}_s\omega_1 \\ \vdots & & & & \vdots & & \\ \sigma_1\omega_n & \cdots & \sigma_r\omega_n & \tau_1\omega_n \cdots \tau_s\omega_n & \overline{\tau}_1\omega_n & \cdots & \overline{\tau}_s\omega_n \end{pmatrix}$$

This number is a (positive or negative) rational integer, <u>nonzero</u> because $K/\mathbb{Q}$ is separable (cf. Cassels, Appendix B). (<u>Exercise</u>. The sign of D_K is $(-1)^s$.)

(d) K as above. If v is infinite, and we complete K in the metric topology of $|\ |_v$, we get $K_v = \mathbb{R}$ (v real) or $\mathbb{C}$(v imaginary). Moreover, $|\ |_v$ extends uniquely to K_v. Similarly, if v is finite we get a completion K_v (written $\mathbb{Q}_p$ for $K = \mathbb{Q}$, v = p-adic valuation). Let O_v = ring of local integers $= \{\alpha \in K_v \mid |\alpha|_v \leq 1\}$ and P_v = unique maximal ideal of $O_v = \{\alpha \in K_v | \ |\alpha_v| < 1\}$. For some $\pi_v \in O_K$ such that $|\pi_v|_v$ is maximal < 1 (π_v is unique up to units), $P_v = \pi_v O_v$. Moreover, $k_v = O_v/P_v$ is <u>finite</u> ($\cong \mathbb{F}_q$ if q is the "norm" of the prime in O_K which defines v). If Σ is a set of coset representatives for k_v, we can express $\alpha \in K_v$ uniquely as a Laurent series in powers of π_v with coefficients in Σ. (O_v consists of the ordinary power series.)

§3. Adele ring of a global field

In the 1930's Chevalley invented ideles (see §7 below); the additive version (adeles, or valuation vectors) is now widely used as well. Essentially, adeles provide a formalism for studying simultaneously all completions K_v (v finite or infinite) of a global field K, cf. Robert [1]. Tate's thesis, for example, exploits the possibilities of this formalism in the direction of Fourier analysis. In this section we merely introduce the basic notions, following Cassels [1, §13-14] (cf. also Tate [1, §3].)

3.1 Restricted topological products

Let X_λ $(\lambda \in \Lambda)$ be a collection of topological spaces, with open subsets Y_λ defined for almost all λ. Let X consist of all elements $(x_\lambda) \in \prod_{\lambda \in \Lambda} X_\lambda$ satisfying: $x_\lambda \in Y_\lambda$ for almost all λ. Topologize X by taking as basic open sets all products $\prod_{\lambda \in \Lambda} Z_\lambda$, where Z_λ is open in X_λ and $Z_\lambda = Y_\lambda$ for almost all λ. Call X (with this topology) the <u>restricted topological product</u> of the X_λ with respect to the Y_λ.

If $S \subset \Lambda$ is finite, and includes all λ for which Y_λ is not defined, let

$$X(S) = \prod_{\lambda \in S} X_\lambda \times \prod_{\lambda \notin S} Y_\lambda$$

with the product topology. A moment's thought shows that X is the union of the various open subsets X(S), and indeed the topology of X is uniquely specified by the requirement that each X(S) (with its product topology) be open in X. (X is the direct limit of the X(S).)

LEMMA. *If X_λ is locally compact, and Y_λ is compact whenever defined, then X is locally compact.*

Proof. By Tychonoff's Theorem, $\prod_{\lambda \notin S} Y_\lambda$ is compact, so the product $X(S)$ is locally compact (as a finite product of locally compact spaces). Since $X(S)$ is open in X, and $X = \bigcup_S X(S)$, each $x \in X$ has a compact neighborhood. □

We also want to discuss measure. If each X_λ has a measure μ_λ, with $\mu_\lambda(Y_\lambda) = 1$ (whenever Y_λ is defined), the product measure on $X(S)$ is well defined. It is easy to see that the various product measures μ_S are compatible and hence define a measure μ on X. A basis for the measurable sets will be all $M = \prod_{\lambda \in \Lambda} M_\lambda$, where $\mu_\lambda(M_\lambda)$ is defined and $M_\lambda = Y_\lambda$ for almost all λ; then $\mu(M) = \prod_{\lambda \in \Lambda} \mu_\lambda(M_\lambda)$.

3.2 Adeles

Let K be a global field. For almost all valuations v, O_v is defined and is a compact open subgroup of the additive locally compact group K_v (§ 2). From 3.1 we conclude readily that the restricted topological product of the K_v with respect to the O_v (when defined), v running over all finite and infinite valuations, is a locally compact topological ring $\mathbb{A}_K$ (when endowed with componentwise operations). The reader should verify that the ring operations are continuous. Similarly, $\mathbb{A}_K(S) = \prod_{v \in S} K_v \times \prod_{v \notin S} O_v$ is defined for finite $S \supset S_\infty$ (the set of archimedean v). Call $\mathbb{A}_K$ the ring of adeles, $\mathbb{A}_K(S)$ the ring of S-adeles of K. Denote $\mathbb{A}_K(S_\infty)$ by $\mathbb{A}_K(\infty)$.

If we normalize Haar measure μ_v on K_v so that $\mu_v(O_v) = 1$ (v finite), which is possible since O_v is compact, 3.1 yields a measure μ on $\mathbb{A}_K$. (Exercise: This is Haar measure on the locally compact group $\mathbb{A}_K$.)

Finally, use the fact (§2, Appendix) that if $\alpha \in K^*$, $|\alpha|_v \leq 1$ for almost all v, to embed K in $\mathbb{A}_K$ as the set of principal adeles $(\alpha, \alpha, \ldots)$. In §4 we will see that K is discrete in $\mathbb{A}_K$.

Exercise (not trivial): Let K be a global field, L/K a finite separable extension (again a global field!). Then $\mathbb{A}_K \otimes_K L$ is canonically isomorphic (topologically and algebraically) to $\mathbb{A}_L$.

II. THE ADDITIVE GROUP

Here our aim is to study the analogue for adeles of the action of $\mathbb{Z}$ on $\mathbb{R}$ by translations (or the action of O_K on the product of the archimedean completions of the number field K). Some of the results will make sense also in the function field case, but for the sake of clarity we assume always that K is a number field. (The reader is urged to fill in details for the function field case whenever possible.) It turns out that the appropriate analogue for the pair $\mathbb{Z}$, $\mathbb{R}$ is the pair $\mathbb{Q}$, $\mathbb{A}_\mathbb{Q}$ (or K, $\mathbb{A}_K$ in general). For the most part we follow Cassels [1, §13-15]. K always denotes a number field, of degree $n = [K:\mathbb{Q}]$.

§4. The quotient $\mathbb{A}_K/K$

4.1 The space K_∞

We follow the notation of the appendix to §2. In particular, $\{\omega_1,\ldots,\omega_n\}$ is an integral basis of $K/\mathbb{Q}$, $O_K = \mathbb{Z}\omega_1 + \ldots + \mathbb{Z}\omega_n$. If the σ_i, τ_i are the various embeddings of K, we write (for convenience)

$$\omega_{ij} = \sigma_j(\omega_i) \qquad j = 1,\ldots,r$$

$$\xi_{ij} + \sqrt{-1}\,\eta_{ij} = \tau_j(\omega_i) \qquad j = 1,\ldots,s .$$

Then by definition of the discriminant D_K ,

$$\sqrt{|D_K|} = \left|\det\begin{pmatrix} \omega_{11} & \cdots & \omega_{1r} & \xi_{11}+\sqrt{-1}\,\eta_{11} & \cdots & \xi_{11}-\sqrt{-1}\,\eta_{11} & \cdots \\ \vdots & & \vdots & \vdots & & \vdots & \\ \omega_{n1} & \cdots & \omega_{nr} & \xi_{n1}+\sqrt{-1}\,\eta_{n1} & \cdots & \xi_{n1}-\sqrt{-1}\,\eta_{n1} & \cdots \end{pmatrix}\right|$$

$$= |(-\sqrt{-1})^s 2^s| \left| \det \begin{pmatrix} \omega_{11} & \cdots & \omega_{1r} & \xi_{11} & \cdots & \xi_{1s} & \eta_{11} & \cdots & \eta_{1s} \\ \vdots & & \vdots & \vdots & & \vdots & \vdots & & \vdots \\ \omega_{n1} & \cdots & \omega_{nr} & \xi_{n1} & \cdots & \xi_{ns} & \eta_{n1} & \cdots & \eta_{ns} \end{pmatrix} \right|$$

(by elementary column operations).

Since $D_K \neq 0$, the rows of this last matrix (with real entries) must be linearly independent over $\mathbb{R}$. For the moment this is all the information we need; in §5 we shall return to the above calculation, however.

Now define $K_\infty = \prod_{v \text{ infinite}} K_v$. (For $K = \mathbb{Q}$, this is just $\mathbb{R}$.) Recall what K_v looks like: v is either real or imaginary, associated with an embedding $\sigma_i : K \to \mathbb{R}$ or a pair $\tau_i, \bar{\tau}_i : K \to \mathbb{C}$. In the former case, K_v is identifiable with $\mathbb{R}$ if we replace $\omega \in K$ by $\sigma_i(\omega)$ and form all Cauchy sequences in the latter elements. Moreover, $|\omega|_v$ is the usual absolute value of $\sigma_i(\omega)$. The imaginary case is similar, except that $|\omega|_v$ is the square of the usual absolute value of $\tau_i(\omega)$ (or $\bar{\tau}_i(\omega)$!). (Recall how we normalized valuations.)

Therefore we may identify K_∞ with $\underbrace{\mathbb{R} \times \ldots \times \mathbb{R}}_{r} \times \underbrace{\mathbb{C} \times \ldots \times \mathbb{C}}_{s} \cong \mathbb{R}^n$; if we take $\omega \in K$, it becomes identified with $(\sigma_1(\omega), \ldots, \sigma_r(\omega), \tau_1(\omega), \ldots, \tau_s(\omega))$ or with the vector $(\sigma_1(\omega), \ldots, \sigma_r(\omega), \xi_1(\omega), \eta_1(\omega), \ldots, \xi_s(\omega), \eta_s(\omega))$ in $\mathbb{R}^n$ if we write $\tau_i(\omega) = \xi_i(\omega) + \sqrt{-1}\,\eta_i(\omega)$. In particular, the elements of $\mathbb{R}^n$ corresponding to our integral basis $\{\omega_1, \ldots, \omega_n\}$ are just the rows of the matrix above (with different order of subscripts), which we found to be linearly independent over $\mathbb{R}$. Moreover, $O_K = \mathbb{Z}\omega_1 + \ldots + \mathbb{Z}\omega_n$ becomes a lattice in $\mathbb{R}^n$ (i.e., the $\mathbb{Z}$-span of a basis of $\mathbb{R}^n$), obviously a discrete subgroup in the usual topology, which corresponds exactly to the product topology on K_∞.

Given such a lattice in $\mathbb{R}^n$, there is an obvious choice of fundamental domain:

$$F_\infty = \{ \sum_{i=1}^{n} t_i\omega_i \in K_\infty \mid 0 \leq t_i < 1\} .$$

4.2 Fundamental domain for K in $\mathbb{A}_K$

LEMMA. K is a discrete subgroup of $\mathbb{A}_K$.

Proof. Suppose not. Then each neighborhood of zero must meet K^*, for instance any neighborhood $M \times \prod_{v \text{ finite}} O_v$ (M = neighborhood of zero in K_∞). But $\alpha \in O_v$ for all finite v iff $|\alpha|_v \leq 1$ (v finite) iff $\alpha \in O_K$, and we saw in 4.1 that O_K is discrete in K_∞. Contradiction. □

THEOREM. Let $F = F_\infty \times \prod_{v \text{ finite}} O_v$. Then F is a fundamental domain for K in $\mathbb{A}_K$. F is relatively compact (i.e., has compact closure) and thus $\mathbb{A}_K/K$ is compact in the quotient topology.

Proof. The second assertion is obvious once the first is proved. We notice that F is included in $\mathbb{A}_K(\infty) = K_\infty \times \prod_{v \text{ finite}} O_v$, so as a first step we show that any adele can be translated by K into $\mathbb{A}_K(\infty)$: $\mathbb{A}_K = K + \mathbb{A}_K(\infty)$. Let α be any adele. Only finitely many $|\alpha_v|_v > 1$, so clearly there exists a big enough $m \in \mathbb{Z}$ so that

$$|m\alpha_v|_v \leq 1 \qquad (v \text{ finite}).$$

(If $|\alpha_v|_v \leq 1$ already, multiplying by an integer doesn't reverse the inequality.) The set S of finite v for which $|m|_v \neq 1$ is finite, so we may use a version of the Chinese Remainder Theorem (see 6.1 below) to find $\beta \in K$ satisfying:

$$|m\alpha_v - \beta|_v \leq |m|_v \qquad (v \in S)$$

$$|\beta|_v \leq 1 \qquad (v \notin S,\ v \text{ finite}).$$

Then $\beta/m \in K$, and $|\alpha_v - \frac{\beta}{m}|_v = |\frac{1}{m}|_v |m\alpha_v - \beta|_v \leq 1$ for all finite v.

Once inside $\mathbb{A}_K(\infty)$, the discussion in 4.1 allows us to translate further by an element of O_K to get inside F. (The only point to check here is that translation by O_K takes $\mathbb{A}_K(\infty)$ into itself.)

Finally, if two distinct elements $\alpha, \alpha' \in F$ differ by an element of K, say β, then

$$|\beta|_v = |\alpha_v - \alpha'_v|_v \leq \max(|\alpha_v|_v, |\alpha'_v|_v) \leq 1 \quad \text{for finite } v,$$

so that $\beta \in O_K, \beta \neq 0$. This contradicts the fact that F_∞ is a fundamental domain for O_K in K_∞. Therefore F meets each coset of $\mathbb{A}_K/K$ exactly once.□

Remark. If K is allowed to be a function field, then it is still true that K is discrete in $\mathbb{A}_K$ and $\mathbb{A}_K/K$ is compact.

4.3 Product formula

Recall that we normalized our valuations using the notion of module of an automorphism (2.1). That this normalization has good properties is borne out by the product formula, which is obvious for $\mathbb{Q}$ and not hard to verify directly for arbitrary K, but which Tate [1] obtained cleverly in the context of adeles. This will play a vital role in the discussion of ideles. One preliminary lemma is needed (and will be needed again later).

LEMMA. *Let μ be a Haar measure on the locally compact additive group $\mathbb{A}_K$, F as above. If $E \subset \mathbb{A}_K$ is μ-measurable and maps injectively to $\mathbb{A}_K/K$ (under the canonical map), then $\mu(E) \leq \mu(F)$. If E also maps onto $\mathbb{A}_K/K$, then $\mu(E) = \mu(F)$.*

Proof. The only property of F we need is that it should be a μ-measurable fundamental domain; therefore, the second statement follows from the first by interchanging E and F. To prove the first statement, notice that by definition of fundamental domain the distinct K-translates of F are disjoint and cover $\mathbb{A}_K$. So the intersections $(\beta+F) \cap E$ $(\beta \in K)$ are disjoint and cover E, and these are obviously μ-measurable. But $(\beta+F) \cap E$ is a K-translate of $(-\beta+E) \cap F$ (draw a picture!) and hence has the same Haar measure as the latter subset of F. These subsets of F are disjoint, thanks to the hypothesis on E, so the sum of their measures, namely $\mu(E)$, cannot exceed $\mu(F)$.□

PRODUCT FORMULA: If $\alpha \in K^*$, $\prod_v |\alpha|_v = 1$.

Proof. By definition, multiplication by α in K_v changes Haar measure there by a factor $|\alpha|_v$, so (clearly) multiplication by α in $\mathbb{A}_K$ magnifies Haar measure by $\prod_v |\alpha|_v$. But αF is again a fundamental domain for K in $\mathbb{A}_K$ (because $\alpha K = K$), so the preceding lemma shows that its Haar measure equals $\mu(F)$ (and is finite because F is compact). The conclusion follows. □

§5. Volume of fundamental domain

5.1 Normalized Haar measure

As we pointed out in 3.2, the choice of Haar measure μ_v in each K_v, such that $\mu_v(O_v) = 1$ for all (or almost all) finite v, determines a Haar measure μ in $\mathbb{A}_K$. The restriction of μ to $\mathbb{A}_K(S)$ is just the product measure. Here we make specific choices as follows: For finite v, require $\mu_v(O_v) = 1$ (this determines μ_v uniquely). For real v, let μ_v be ordinary Lebesgue measure in $\mathbb{R}$, and for imaginary v, take twice the usual Lebesgue measure in the complex plane. (These latter choices are dictated by the desire to make μ_v self-dual relative to the Fourier transform: see Tate [1,2.2]. To be consistent we ought to require $\mu_v(O_v) = (N\partial_v)^{-1/2}$ for finite v (∂_v = "local different"), which would yield $\mu(F) = 1$ below, but we shall avoid discussing differents.)

Exercise.

(a) If v is real, $\{\alpha_v \in K_v \mid |\alpha_v|_v \leq C\} = T$ is an interval and $\mu(T) = 2C$.

(b) If v is imaginary, the set T in (a) is a disc in the complex plane and $\mu_v(T) = 2\pi C$. [Recall that $|\ |_v$ is the square of the usual absolute value in $\mathbb{C}$!]

(c) If v is finite, T as above, then $\mu_v(T) \leq C$, with equality iff C is in the value group (the infinite cyclic group generated by $1/\text{Card } k_v$ in $\mathbb{R}^{>0}$). [Hint: Count cosets.]

5.2 Volume calculation

THEOREM. Let μ be Haar measure in $\mathbb{A}_K$, normalized as above. Then if F is the fundamental domain constructed in §4, $\mu(F) = \sqrt{|D_K|}$.

Proof. First of all, it is obvious that $\mu(F) = \text{vol}\,(F_\infty)$, where volume in K_∞ is computed using the product measure $\prod \mu_v$ (v infinite). But F_∞ is just a parallelotope in $\mathbb{R}^n$ determined by the vectors $(\omega_{i1},\ldots,\omega_{ir},\xi_{i1},\eta_{i1},\ldots,\xi_{is},\eta_{is})$, $1 \leq i \leq n$. As is well known, the volume of such a parallelotope is given (up to sign) by the determinant of the matrix having these vectors as its rows. We computed this determinant in 4.1: its absolute value is $\sqrt{|D_K|}$, apart from a factor 2^s which disappears because of our normalization of Haar measure. □

Remarks. (a) In view of Lemma 4.3, any measurable fundamental domain will have measure $\mu(F)$.

(b) It is not easy to describe F directly by inequalities of the form

$$|\alpha_v|_v \leq \delta_v \qquad (v \text{ infinite})$$

$$|\alpha_v|_v \leq 1 \qquad (v \text{ finite}).$$

However, by choosing the δ_v to be large enough, it is clear that we can force this (compact) set to include F; its measure (computable readily by the exercise in 5.1) will be $\geq \mu(F)$.

5.3 Application: Fields of discriminant ± 1

The following classical result is not hard to deduce; we follow the proof in Weil [2, p.92], recalling first that $|D_K|$ is a positive (rational) integer and that $D_{\mathbb{Q}} = 1$.

THEOREM. If $|D_K| = 1$, then $K = \mathbb{Q}$. (Equivalently, $K \neq \mathbb{Q}$ implies $|D_K| > 1$.)

Proof. Label the archimedean valuations

$$\underbrace{|\ |_1, \ldots, |\ |_r}_{\text{real}} \qquad \text{and} \qquad \underbrace{|\ |_{r+1}, \ldots, |\ |_{r+s}}_{\text{imaginary}} ,$$

recalling that $[K:\mathbb{Q}] = n = r+2s$.

If $c = (c_1, \ldots, c_{r+s})$ are $r+s$ positive reals, we let $F(c)$ be the compact subset of all $\alpha \in \mathbb{A}_K$ satisfying

$$|\alpha_i|_i \leq \frac{c_i}{2} , \quad v_i \text{ infinite}$$

$$|\alpha_v|_v \leq 1 \quad , \quad v \text{ finite} .$$

By the exercise in 5.1, $\mu(F(c)) = \pi^s (\prod_i c_i)$. By Lemma 4.3, $\mu(F(c)) > \mu(F)$ implies that $F(c) \to \mathbb{A}_K/K$ is not injective, i.e., there exist $\gamma, \gamma' \in F(c)$ such that $\beta = \gamma - \gamma' \in K^*$. This clearly forces:

$$|\beta|_i \leq c_i \quad , \quad v_i \text{ real}$$

$$|\beta|_i \leq 2c_i \ , \quad v_i \text{ imaginary}$$

$$|\beta|_v \leq 1 \quad , \quad v \text{ finite} .$$

But then the Product Formula (4.3) requires $2^s \prod_i c_i \geq \prod_v |\beta|_v = 1$, since β is nonzero.

With these preliminaries disposed of, we can prove the theorem. Let $|D_K| = 1$; we have to show that $n = 1$. Of course, the hypothesis means that $\mu(F) = 1$ (Theorem 5.2).

(1) Suppose $s > 0$. Then it is possible to choose the $c_i > 0$ so that $\pi^{-s} < \prod_i c_i < 2^{-s}$. This forces $\mu(F(c)) > 1$, whence we get $2^s \prod_i c_i \geq 1$ (by the above remarks), a contradiction.

(2) Now we have $s = 0$, $r = n$. Since for $c_i = 4$ $(1 \leq i \leq r+s)$, $F(c)$ is compact, it contains only finitely many $\beta \in K$ (K being discrete). Therefore we may choose $1 < c' < 2$ such that none of these satisfies $1 < |\beta|_1 < c'$. Next choose c_1 with $1 < c_1 < c'$,

and if $n > 1$ choose $c_i < 1$ $(i > 1)$ so close to 1 that $\prod_i c_i > 1$. As before, $\mu(F(c)) > \mu(F) = 1$, so there exists $\beta \in K^*$ such that

$$|\beta|_i \le c_i < 2 \quad (v_i \text{ infinite})$$
$$|\beta|_v \le 1 \quad (v \text{ finite}).$$

By choice of c_1, since β lies in $F(4)$, we have $|\beta|_i \le 1$. But $c_i < 1$ for $i > 1$, so $|\beta|_i < 1$ for $i > 1$. If $n > 1$ this contradicts the Product Formula. Therefore $n = 1$.□

Exercise. Work out details of the proof of Corollary 2 in Weil [2, p.92]: There exist only finitely many extensions K of $\mathbb{Q}$ of degree n having a given discriminant.

§6. Strong approximation

The theorem to be proved here is the prototype of a theorem valid for a large class of linear groups, cf. 14.3 below.

6.1 Chinese Remainder Theorem

Starting with the familiar Chinese Remainder Theorem for $K = \mathbb{Q}$, we list some equivalent versions.

(1) Given $\alpha_1,\dots,\alpha_t \in \mathbb{Z}$, pairwise relatively prime integers $n_i > 0$, there exists $\alpha \in \mathbb{Z}$ such that $\alpha \equiv \alpha_i \pmod{n_i}$ for all i.

(2) Same as (1), but letting the n_i be powers of distinct primes only.

(3) Given $\alpha_1,\dots,\alpha_t \in \mathbb{Z}$, distinct primes $p_1,\dots,p_t$, and $\varepsilon_1,\dots,\varepsilon_t > 0$, there exists $\alpha \in \mathbb{Z}$ such that

$$|\alpha-\alpha_i|_{p_i} < \varepsilon_i \quad (\text{all } i).$$

(4) Given $\alpha_1,\dots,\alpha_t \in \mathbb{Q}$, distinct primes $p_1,\dots,p_t$, and $\varepsilon_i > 0$, there exists $\alpha \in \mathbb{Q}$ such that

$$|\alpha-\alpha_i|_{p_i} < \varepsilon_i \quad (\text{all } i)$$
$$|\alpha|_p \le 1 \quad (p \text{ prime} \ne p_1,\dots,p_t)$$

(5) Same as (4), but allow $\alpha_i \in \mathbb{Q}_{p_i}$.

The proofs that these are equivalent are mostly obvious. To get (4) => (5), just use the fact that each $\alpha_i \in \mathbb{Q}_{p_i}$ can first be approximated by an element of $\mathbb{Q}$ to within $\varepsilon_i/2$. (3) => (4) Let $m \in \mathbb{Z}$ be a common denominator: $\alpha_i = \frac{\beta_i}{m}$ $(\beta_i \in \mathbb{Z})$. Let q_i be the primes $\neq p_j$ dividing m. Apply (3) to find $\beta \in \mathbb{Z}$ with

$$|\beta-\beta_i|_{p_i} < |m|_{p_i} \quad \varepsilon_i$$
$$|\beta|_{q_i} \leq |m|_{q_i}$$

Set $\alpha = \frac{\beta}{m} \in \mathbb{Q}$, and check that α does what is required.

For an arbitrary number field K, the Chinese Remainder Theorem takes the following form (see Goldstein [1, Thm. 2-2-13]): Given a finite set S of nonarchimedean valuations, with $\alpha_v \in K_v$ and $\varepsilon_v > 0$ $(v \in S)$, there exists $\alpha \in K$ such that $|\alpha-\alpha_v|_v < \varepsilon_v$ $(v \in S)$ while $|\alpha|_v \leq 1$ $(v \notin S$, v nonarchimedean). We used this already in the proof of Theorem 4.2.

6.2 An important lemma

LEMMA. *There is a constant* $C > 0$ *(depending only on* K) *such that if* $\alpha \in \mathbb{A}_K$, $\prod_v |\alpha_v|_v > C$, *then there exists* $\beta \in K^*$ *satisfying* $|\beta|_v \leq |\alpha_v|_v$ *for all* v.

Proof. (We remark that the product $\prod_v |\alpha_v|_v$ is actually finite: almost all $|\alpha_v|_v \leq 1$, and if almost all $|\alpha_v|_v < 1$, it is easy to see that the product would tend to 0. The reader can look ahead to §7 for more details.)

Take for C the Haar measure of a fundamental domain for K in A_K (§5). If α satisfies the hypothesis of the lemma, let T be the set of all adeles γ satisfying:

$$|\gamma_v|_v \leq \begin{cases} \dfrac{|\alpha_v|_v}{2} & v \text{ real} \\ \dfrac{|\alpha_v|_v}{2\pi} & v \text{ imaginary} \\ |\alpha_v|_v & v \text{ finite} \end{cases}.$$

According to the exercise in 5.1, $\mu(T) = \prod_v |\alpha_v|_v > C$. Then Lemma 4.3 implies that there exist $\gamma, \gamma' \in T$ with $\gamma - \gamma' = \beta \in K^*$, whence

$$|\beta|_v = |\gamma-\gamma'|_v \leq \begin{cases} |\alpha_v|_v & v \text{ real} \\ \frac{2}{\pi} |\alpha_v|_v & v \text{ imaginary} \\ |\alpha_v|_v & v \text{ finite.} \end{cases} \qquad \square$$

COROLLARY. *Let v_0 be an arbitrary valuation of K. Given $\delta_v > 0$ $(v \neq v_0)$, with $\delta_v = 1$ for almost all v, there exists $\beta \in K^*$ such that $|\beta|_v \leq \delta_v$ $(v \neq v_0)$. (Intuitively, this says that we can find a nonzero element of K in an arbitrarily small box around 0, provided one dimension of the box is unrestricted.)*

Proof. Set $\alpha_v = 1$ whenever $\delta_v = 1$ and find α_v such that $0 < |\alpha_v|_v \leq \delta_v$ otherwise (clearly possible). Then choose $\alpha_{v_0} \in K_{v_0}$ so that

$$\prod_v |\alpha_v|_v > C .$$

According to the lemma, our desired $\beta \in K^*$ exists. $\square$

6.3 Main theorem

STRONG APPROXIMATION THEOREM. *Fix some v_0. Let B be the restricted product (3.1) of the K_v with respect to the $\mathcal{O}_v (v \neq v_0)$, and embed K in B (as we did in $\mathbb{A}_K$). Then K is dense in B (whereas it was discrete in $\mathbb{A}_K$).*

Proof. (M. Kneser) We must show that each open set in B meets K. Any open set contains one of the form: set of $\gamma \in B$ such that

$$|\gamma_v - \alpha_v|_v < \varepsilon \qquad (v \in S)$$
$$|\gamma_v|_v \leq 1 \qquad (v \notin S,\ v \neq v_0) ,$$

where S is a finite set $\supset S_\infty - \{v_0\}$ and $\varepsilon > 0$, $\alpha_v \in K_v$.

As remarked in 5.2, some set $E \subset \mathbb{A}_K$ defined by inequalities $|\gamma_v|_v \leq \delta_v$ ($\delta_v = 1$ for almost all v, including all finite v)

contains a fundamental domain for K in $\mathbb{A}_K$. Use the corollary to Lemma 6.2 to find $\lambda \in K^*$ satisfying

$$|\lambda|_v < \varepsilon/\delta_v \qquad (v \in S)$$
$$\leq 1/\delta_v \qquad (v \notin S,\ v \neq v_0) \quad .$$

Then λE again contains a fundamental domain, so each $\alpha \in \mathbb{A}_K$ is the sum of an element of λE and an element $\beta \in K$. This can be applied to the adele α whose components are the given α_v $(v \in S)$ or 0 $(v \notin S)$, to yield the desired $\beta \in K$. □

III. THE MULTIPLICATIVE GROUP

As before, K is a number field, of degree $n = r+2s$ over $\mathbb{Q}$. We continue to use the notation of I, following Cassels [1, §16-18].

§7. Ideles

7.1 Idele topology

Let $U_v = \{\alpha \in K^* \mid |\alpha|_v = 1\}$ for finite v, or equivalently $U_v = O_v - P_v$. We call U_v the group of "v-adic units". Because $|\ |_v$ is continuous and the value group is discrete, U_v is open, as well as closed and compact. Therefore it makes sense to form the restricted topological product J_K of the locally compact spaces K_v^* (all v) with respect to the compact open subspaces U_v (v finite) (see 3.1). Endowed with componentwise multiplication, J_K becomes a locally compact group (the reader should check that inversion is continuous, using the fact that it is continuous in each K_v^*). We call J_K the <u>group of ideles</u> of K.

From the viewpoint of the general theory of adelic linear groups this is the appropriate way to approach J_K. But there is a more down-to-earth way. An idele α is just an element of the product $\prod_v K_v^*$ such that $|\alpha_v|_v = 1$ for almost all v, so α may also be regarded as an adele. In this way the <u>set</u> J_K is obviously identified with the group of units of the ring $\mathbb{A}_K$ (to be invertible, an adele α must have all its components nonzero and must satisfy $|\alpha_v|_v \le 1$, $|\alpha_v|_v^{-1} \le 1$ for almost all v).

Unfortunately, the relative topology for the subset J_K of $\mathbb{A}_K$ fails to make inversion continuous (check this). We can overcome this problem by throwing in more open sets: topologize J_K as the subset of $\mathbb{A}_K \times \mathbb{A}_K$ consisting of all pairs (α,α^{-1}). (If $\delta(\alpha) = (\alpha,\alpha^{-1})$, then U open in $\mathbb{A}_K$ implies $U \times \mathbb{A}_K$ open in $\mathbb{A}_K \times \mathbb{A}_K$, hence $\delta^{-1}(U \times \mathbb{A}_K) = U \cap J_K$ is open in this new topology on J_K.)

<u>Claim</u>. This topology on J_K coincides with the restricted product topology.

This depends on the fact that inversion is continuous in each K_v^*. In one direction, take a basic open set in $\mathbb{A}_K \times \mathbb{A}_K$; it contains an open set $(\prod_{v\in S} M_v \times \prod_{v\notin S} O_v) \times (\prod_{v\in S} N_v \times \prod_{v\notin S} O_v)$, with $S \supset S_\infty$ and M_v, N_v open in K_v. δ^{-1} gives an open set in J_K (in the new topology) of the form $\{\alpha\in J_K \,|\, \alpha_v \in M_v \cap N_v^{-1} \cap K_v^*,\ v \in S\}$, which is open in the restricted product topology. The other direction is similar. □

Therefore we are free to view J_K as a subset of $\mathbb{A}_K$, with the above refinement of the relative adele topology. K^* embeds naturally in J_K.

<u>LEMMA</u>. <u>K^* is discrete in J_K</u>.

<u>Proof</u>. K (hence K^*) is a discrete subset of $\mathbb{A}_K$ (4.2), so $K^* \times K^*$ is discrete in $\mathbb{A}_K \times \mathbb{A}_K$ and $\delta^{-1}(K^* \times K^*)$ is discrete in J_K. □

<u>Exercise</u>. Formulate a "strong approximation" theorem for J_K and decide whether it is valid.

7.2 <u>Special ideles</u>

Let $c(\alpha) = \prod_v |\alpha_v|_v$ $(\alpha \in J_K)$. Since $|\alpha_v|_v = 1$ for almost all v, this is actually a finite product, called the <u>content</u> (or "volume") of α. The map $c : J_K \to \mathbb{R}^{>0}$ is evidently a continuous homorphism (continuous essentially because each $|\ |_v$ is continuous). Its kernel J_K^0 is a closed subgroup of J_K, which we call the group of special ideles. (There is a rough analogy with GL_n, SL_n, det, which is brought out in the exercise below.)

LEMMA. $K^* \subset J_K^0$.

Proof. Product Formula (4.3).□

Exercise. Define $c : \mathbb{A}_K \to \mathbb{R}^{\geq 0}$ by

$$c(\alpha) = \prod_v |\alpha_v|_v$$

(a) This is a well defined map.

(b) $c(\alpha) \neq 0$ iff $\alpha \in J_K$. [Hint: Try the case $K = \mathbb{Q}$ first.]

(c) Topologize $\mathbb{R}^{\geq 0}$ as follows: the nonempty open sets are to be the sets $V \cup \{0\}$, where V is open in the usual (subspace) topology on $\mathbb{R}^{\geq 0} \subset \mathbb{R}$. Check that this is a topology. Which points are closed?

(d) For the topology just given to $\mathbb{R}^{\geq 0}$, c is continuous.

(e) c is not continuous if $\mathbb{R}^{\geq 0}$ has its usual topology. [Hint: Notice that J_K is not open in $\mathbb{A}_K$.]

(f) As a corollary of (d), deduce that J_K^0 is closed in $\mathbb{A}_K$.

Before starting the next lemma, we recall that for finite v,

$$|\alpha|_v = \left(\frac{1}{\text{Card } k_v}\right)^{\text{ord}_v(\alpha)} \qquad \text{for } \alpha \in K.$$

This normalization was the one imposed on us by Haar measure considerations (2.1, exercise), since $\text{Card } k_v$ = index of P_v in O_v. In particular, if $\alpha_v \in O_v$ but $\alpha_v \notin U_v$, then $|\alpha_v|_v \leq \frac{1}{2}$ for any finite v.

LEMMA. The relative idele and adele topologies on J_K^0 coincide.

COROLLARY. J_K^0 is closed in $\mathbb{A}_K$ (cf. preceding exercise).

Proof of Lemma. As remarked earlier, if U is open in $\mathbb{A}_K$, then $U \cap J_K$ is open in the idele topology, so $U \cap J_K^0$ is open in the relative idele topology.

Conversely, let $\alpha \in J_K^0$. Any neighborhood of α in the relative idele topology obviously includes a basic open set $U \subset J_K$ intersected with J_K^0, where U has the special form

$$\{\gamma \in J_K \mid |\gamma_v - \alpha_v|_v < \varepsilon \quad (v \in S), \quad |\gamma_v|_v = 1 \quad (v \notin S)\}$$

and where S includes all finite v with $|\alpha_v|_v \neq 1$. By definition $\prod_v |\alpha_v|_v = \prod_{v \in S} |\alpha_v|_v = 1$, so by requiring ε to be very small we can insure that all $\gamma \in U$ satisfy $\prod_v |\gamma_v|_v = \prod_{v \in S} |\gamma_v|_v < 2$. Now all we need to do is find an open set $U' \subset \mathbb{A}_K$ such that $U' \cap J_K^0 \subset U \cap J_K^0$. Take $U' = \{\gamma \in \mathbb{A}_K \mid |\gamma_v - \alpha_v|_v < \varepsilon (v \in S), |\gamma_v|_v \leq 1 \ (v \notin S)\}$. (So indeed we have the opposite inclusion!) Then $\gamma \in U' \cap J_K^0$ forces $\prod_{v \in S} |\gamma_v|_v < 2$ (by the above choice of ε). Since $\prod_v |\gamma_v|_v = 1$, and since $|\gamma_v|_v \leq 1$ $(v \notin S)$, our remark preceding the lemma shows that $|\gamma_v|_v = 1$ $(v \notin S)$, whence $\gamma \in U$ as required. □

§8. Compactness theorem

8.1 Compactness of J_K^0/K^*

THEOREM. J_K^0/K^* is compact in the quotient topology.

Proof. Because of the lemma and corollary in 7.2, it suffices to find a compact set $W \subset \mathbb{A}_K$ such that $W \cap J_K^0 \to J_K^0/K^*$ is surjective, i.e., such that $J_K^0 = (W \cap J_K^0)K^*$. For this we exploit the existence of a compact fundamental domain for $\mathbb{A}_K/K$ proved in §4. Let ε be any idele of content $> C$ (C as in Lemma 6.2), and set $W = \{\gamma \in \mathbb{A}_K \mid |\gamma_v|_v \leq |\varepsilon_v|_v$ for all $v\}$. Evidently W is compact in the adele topology, and $|\varepsilon_v|_v = 1$ for almost all v. Now if $\alpha \in J_K^0$ is any special idele, then $c(\alpha^{-1}\varepsilon) = c(\alpha^{-1})c(\varepsilon) = c(\varepsilon) > C$, so by Lemma 6.2 there exists $\beta \in K^*$ satisfying $|\beta|_v \leq |\alpha_v^{-1}\varepsilon_v|_v$

(all v), i.e., $\alpha\beta \in W$. □

Exercise. We can embed $\mathbb{R}^{>0}$ in J_K via
$t \mapsto (\underbrace{t^{1/n}, \ldots, t^{1/n}}_{\text{infinite } v}, \underbrace{1, 1, \ldots}_{\text{finite } v})$. Observe that the content of the idele t is then t (look closely at imaginary v!) Prove that $J_K = \mathbb{R}^{>0} \times J_K^0$ (direct product of topological groups); conclude that J_K/K^* is not compact.

8.2 Applications: Class number and units of K

(1) The ring of algebraic integers O_K, viewed as Dedekind domain, enjoys a kind of unique factorization: each nonzero (fractional) ideal of K, or (ordinary) ideal of O_K, has a unique expression as product of prime ideals (negative exponents allowed). Recall that a fractional ideal is an O_K-submodule I of K such that $\alpha I \subset O_K$ for some $\alpha \in K^*$. Formally, we define the ideal group (denoted I_K) of K to be the free abelian group on the finite valuations v (which correspond 1 - 1 with the prime ideals of O_K). Principal ideals such as (α), $\alpha \in K^*$, yield formal sums $\sum_{v \text{ finite}} \mathrm{ord}_v(\alpha)v$, and these give us a subgroup P_K of I_K.

The ideles were invented to embody this formalism along with information about units and archimedean valuations. We define

$$\phi: J_K \to I_K$$
$$\alpha \to \sum_{v \text{ finite}} \mathrm{ord}_v(\alpha_v)v .$$

Notice that $\phi(K^*) = P_K$. ϕ is obviously a group homomorphism; it becomes continuous if we endow I_K with the discrete topology (since each $|\ |_v$ is continuous). Moreover, ϕ is certainly surjective, because v is image of the idele with π_v in v^{th} position, 1 elsewhere. It is even true that $\phi(J_K^0) = I_K$, because we can modify the idele just mentioned at an infinite place to make its content 1.

THEOREM (Minkowski). The ideal class group $\mathcal{I}_K = I_K/P_K$ is finite. (Its cardinality h is called the class number of K.)

Proof. ϕ induces a continuous map of J_K^0/K^* onto I_K (by the above remarks). The image is compact (by 8.1) and also discrete, hence finite. □

Remark. This is a typical way to obtain a qualitative assertion of finiteness. It tells us nothing, of course, about how big h might be!

Exercise. Treat the function field case (Cassels [1,§17]).

Exercise. What is the kernel of ϕ?

(2) Next we consider the group U_K of units of K, i.e. the invertible elements of O_K, or equivalently $\bigcap_{v \text{ finite}} U_v$.

LEMMA. Let $0 < c_1 \leq c_2 < \infty$. Then $Z = \{\alpha \in U_K | \ c_1 \leq |\alpha|_v \leq c_2$ for all infinite $v\}$ is finite.

Proof. $W = \{\alpha \in J_K | \ c_1 \leq |\alpha_v|_v \leq c_2$ (v infinite) and $|\alpha_v|_v = 1$ (v finite)$\}$ is compact (by Tychonoff), so $W \cap K^* = Z$ is compact and discrete -- hence finite. □

COROLLARY. $\{x \in K| \ |\alpha|_v = 1$ for all $v\}$ is finite, hence equal to the (cyclic) group μ_K of roots of unity in K.

Proof. Take $c_1 = c_2 = 1$. The set Z then clearly forms a group, and is finite by the lemma, so it consists of roots of unity. Conversely, $\mu_K \subset Z$. □

So we see that $\alpha \in K^*$ is a unit iff $|\alpha|_v = 1$ for all finite v, and α is a root of unity iff $|\alpha|_v = 1$ for all v.

To study U_K more closely, notice that $U_K = K^* \cap J_K^0(\infty)$, where $J_K(\infty) = \prod_{v \in S_\infty} K_v^* \times \prod_{v \text{ finite}} U_v$ and $J_K^0(\infty) = J_K^0 \cap J_K(\infty)$. Define $\lambda : J_K(\infty) \to \mathbb{R}^{r+s}$ by $\alpha \mapsto (\log |\alpha_1|_1, \ldots, \log |\alpha_{r+s}|_{r+s})$, where the notation indicates as usual the $r+s$ infinite valuations. Obviously λ is a homomorphism (multiplicative to additive), surjective (because each $| \ |_i$ has image $\mathbb{R}^{>0}$, and continuous (because $| \ |_i$ and log are continuous). λ maps $J_K^0(\infty)$ onto the hyperplane

$X_1+\ldots+X_{r+s} = 0$ in $\mathbb{R}^{r+s}$, which we may identify with $\mathbb{R}^{r+s-1}$. Since λ is open (check!), $\lambda(U_K)$ is a discrete subgroup of $\mathbb{R}^{r+s-1}$.

Exercise. A discrete subgroup of $\mathbb{R}^m$ is finitely generated (hence free) of rank $\leq m$. (Cf. Lang [1, p.107].)

Finally, we notice that $J_K^0(\infty)$ open in J_K^0 implies that $J_K^0(\infty)/U_K = J_K^0(\infty)/K^* \cap J_K^0(\infty)$ is isomorphic to an open (hence closed) subgroup of J_K^0/K^*.

THEOREM (Dirichlet). U_K is finitely generated of rank $r+s-1$, with torsion subgroup μ_K.

Proof. Since J_K^0/K^* is compact (8.1), the preceding remark shows that $J_K^0(\infty)/U_K$ is also compact. Therefore $\lambda(J_K^0(\infty))/\lambda(U_K) = \mathbb{R}^{r+s-1}/\lambda(U_K)$ is compact. This in turn forces $\mathbb{R}^{r+s-1}$/(subspace spanned by $\lambda(U_K)$) to be compact, so this subspace must equal $\mathbb{R}^{r+s-1}$. We know $\lambda(U_K)$ is discrete, so by the exercise above, $\lambda(U_K)$ is free of rank $\leq r+s-1$; since $\lambda(U_K)$ spans $\mathbb{R}^{r+s-1}$, this rank $= r+s-1$. But Ker $(\lambda|U_K) = \mu_K$, obviously (by our lemma), so U_K has the form $\mu_K \times \mathbb{Z}^{r+s-1}$. □

Exercise. If $S \supset S_\infty$ is finite, define the group of S-units $U_K(S)$ to be $\{\alpha \in K^* \mid |\alpha|_v = 1 \text{ for all } v \notin S\}$. (This equals U_K when $S = S_\infty$.) Formulate and prove a suitable analogue of Dirichlet's theorem for S-units.

(3) To round out the above discussion, we observe that, conversely, the finiteness of class number and the structure theorem for U_K together imply the compactness of J_K^0/K^*. Notice first that Ker $\Phi = J_K(\infty)$, so we get an exact sequence of locally compact groups and (continuous) homomorphisms:

$$1 \to \frac{J_K^0(\infty)}{U_K} \to \frac{J_K^0}{K^*} \to \frac{I_K}{P_K} = \mathcal{I}_K \to 0 \quad .$$

By assumption $\mathcal{I}_K$ is finite. On the other hand, the structure theorem for U_K shows that $\lambda(U_K)$ is a lattice in $\mathbb{R}^{r+s-1}$, so the left term of the exact sequence is compact. It follows that the middle term is compact.

8.3 Fundamental domain

We proved the compactness theorem (8.1) without explicitly constructing a fundamental domain for K^* in J_K^0. But it is not hard now to see how the construction should go. We just sketch it, referring the reader to Lang [1, VII, §3] for details.

Let $\alpha \in J_K^0$ be arbitrary. Unlike the case of adeles, we cannot translate α into the "infinite" part $J_K^0(\infty)$ by an element of K^*. Indeed, the discrepancy between $K^*J_K^0(\infty)$ and J_K^0 is precisely the ideal class group I_K. But this is finite, and at the end of the construction we can take translates of the set constructed (one for each coset of P_K in I_K) to get a fundamental domain. Therefore we begin with $\alpha \in J_K^0(\infty)$, and adjust by a unit of K^* to get $\lambda(\alpha)$ into the obvious fundamental domain (parallelotope) for $\lambda(U_K)$ in $\mathbb{R}^{r+s-1}$. This is all uniquely determined up to elements of $\mathrm{Ker}\ (\lambda|U_K) = \mu_K$, so finally we normalize the choice by restricting the "argument" to a fundamental domain on the circle of radius 1 in $\mathbb{C}^*$ (relative to a primitive root of 1 generating μ_K).

The above discussion is imprecise, but it does indicate how several fundamental domains (e.g., for a lattice in $\mathbb{R}^m$) enter into the picture. Moreover, it is clear that the volume of a fundamental domain for K^* in J_K^0 will depend on certain invariants of K: r, s, Card (μ_K), h = class number, D_K, and also the "regulator" R_K, which measures the parallelotope $\mathbb{R}^{r+s-1}/\lambda(U_K)$. See Tate [1, 4.3.2] for details.

IV. GL_n AND SL_n (OVER $\mathbb{R}$)

Following for the most part the discussion in Borel [5] we shall look at "fundamental sets" (approximations to fundamental domains) for $GL(n,\mathbb{Z})$ in $GL(n,\mathbb{R})$, and interpret the results with reference to reduction of quadratic forms. Along the way we introduce Iwasawa and Bruhat decompositions of $GL(n,\mathbb{R})$ (which generalize to other reductive algebraic groups) and adapt everything to $SL(n,\mathbb{R})$ as well.

§9. Example: The modular group

In this section only, G denotes $SL(2,\mathbb{R})$,

$$K = SO(2,\mathbb{R}) = \left\{ \begin{pmatrix} a & b \\ -b & a \end{pmatrix}, a^2 + b^2 = 1 \right\},$$

$$\Gamma = SL(2,\mathbb{Z}) \qquad \text{(the "modular group")} \quad .$$

Let $P = \{z \in \mathbb{C} \mid \operatorname{Im} z > 0\}$ be the Poincaré upper half-plane. G acts on P (on the left) by $\begin{pmatrix} a & b \\ c & d \end{pmatrix} : z \mapsto \frac{az+b}{cz+d}$. This action is not quite effective: its kernel is $\pm I$. It is an easy exercise to check that the action is transitive (for example, compute the orbit of i). The isotropy group of $i = \{\begin{pmatrix} a & b \\ c & d \end{pmatrix} \mid \frac{ai+b}{ci+d} = i\} = K$, clearly. Therefore, we may identify P with the homogeneous space G/K. (The reader should check that the natural topology on P coincides with the quotient topology on G/K).

Following Gel'fand, Graev, Pyatetskii-Shapiro [1] we call $\mathcal{D} \subset P$ a <u>fundamental domain</u> for Γ in P if

(1) $\mathcal{D}$ is open and $\Gamma\overline{\mathcal{D}} = P$;

(2) $\gamma \neq \gamma'$ in $\Gamma/\{\pm I\}$ implies $\gamma\mathcal{D} \cap \gamma'\overline{\mathcal{D}} = \emptyset$ ($\overline{\mathcal{D}}$ = closure of $\mathcal{D}$).

Other definitions are possible, but this will do for now. Notice that we have to be a bit careful about $\pm I$, but this could be avoided by passing to $PSL(2,\mathbb{R})$.

We recall next the classical construction of a fundamental domain for Γ in P . This is well known, and the calculation is done in great detail in Gel'fand, Graev, Pyatetskii-Shapiro [1, Appendix to Chapter 1]. (For another version, cf. Serre [4,VII,§1].)

First, a geometric observation about lattices in the complex plane. We denote Euclidean length by $|z|$, as usual. Let w_1, w_2

span a lattice in the complex plane (this just means that w_1, w_2 are linearly independent), and let $L = \mathbb{Z}w_1 + \mathbb{Z}w_2$. Then if $|w_2| \geq |w_1|$, $|w_2+w_1| \geq |w_2|$, and $|w_2-w_1| \geq |w_2|$, we claim that no nonzero vector in L independent of w_1 has length $< |w_2|$. (The converse of this statement is obvious, since w_2+w_1 and w_2-w_1 are in L.) To prove the claim, draw a circle of radius $|w_2|$ centered at the origin of L; remember that w_2+w_1 and w_2-w_1 are the diagonals of the parallelogram determined by w_1, w_2.

Our candidate for fundamental domain for Γ in P is

$$\mathcal{D} = \{z \in P \mid |z| > 1 \text{ and } |\mathrm{Re}\, z| < \tfrac{1}{2}\}$$

$$= \{z \in P \mid |z| > 1,\ |z+1| > |z|,\ |z-1| > |z|\}$$

(Sketch this region!)

<u>THEOREM.</u> $\mathcal{D}$ <u>is a fundamental domain for</u> Γ <u>in</u> P.

<u>Proof.</u>

(1) First let $z \in P$ be arbitrary. We must find $\gamma \in \Gamma$ such that $\gamma z \in \overline{\mathcal{D}}$. Since $1 \notin P$, 1 and z span the complex plane. If $L = \mathbb{Z}\cdot 1 + \mathbb{Z}\cdot z$, choose nonzero $w_1 = m_{12}z + m_{22}$ in L closest to the origin. Among those vectors in L independent of w_1 choose one as close as possible to the origin and call it $w_2 = m_{11}z + m_{21}$. Set $\gamma = \begin{pmatrix} m_{11} & m_{21} \\ m_{12} & m_{22} \end{pmatrix}$. Clearly $\det \gamma \neq 0$, and by replacing w_1 by $-w_1$ if necessary we may assume $\det \gamma > 0$. By construction, the triangle with vertices $0, w_1, w_2$ contains no other point of L; if $w \in L$ were an interior point of the triangle with vertices w_1, w_2, w_1+w_2, then $w_1+w_2-w \in L$ would be an interior point of the first triangle, which is absurd. We conclude that the parallelogram P determined by w_1, w_2 has no point of L in its interior, hence P is contained in a fundamental domain for the (discrete) group L in $\mathbb{C} = \mathbb{R}^2$. But the latter domain has volume c equal to the area of the parallelogram determined by z, 1; transformation by the matrix

γ multiplies this by $\det\gamma$, so we conclude $\det\gamma \le 1$. Therefore $\det\gamma = 1$ and $\gamma \in \Gamma$. (Cf. 4.3 and 5.2.) Set $z' = \gamma z = w_2/w_1 \in P$. The preceding discussion makes it clear that $|w_2| \ge |w_1|$, $|w_2+w_1| \ge |w_2|$, and $|w_2-w_1| \ge |w_2|$. Dividing all of these by $|w_1|$ yields the defining inequalities for $\bar{\mathcal{D}}$, so $z' \in \bar{\mathcal{D}}$ as required.

(2) Next one has to show that if $\gamma \ne \pm\gamma'$, $\gamma\mathcal{D} \cap \gamma'\bar{\mathcal{D}} = \emptyset$, or equivalently, $\gamma \ne \pm I$ implies $\gamma\mathcal{D} \cap \bar{\mathcal{D}} = \emptyset$. Suppose on the contrary that $z \in \mathcal{D}$, $\gamma = \begin{pmatrix} m_{11} & m_{21} \\ m_{12} & m_{22} \end{pmatrix} \ne \pm I$, and $\gamma z = \frac{m_{11}z+m_{21}}{m_{12}z+m_{22}} = \frac{w_2}{w_1} \in \bar{\mathcal{D}}$. This clearly forces $|w_2| \ge |w_1|$, $|w_2+w_1| \ge |w_2|$, $|w_2-w_1| \ge |w_2|$. By our earlier geometric observation, no nonzero vector in $L = \mathbb{Z}w_1 + \mathbb{Z}w_2$ independent of w_1 has length $< |w_2|$. But since $\gamma \in \Gamma$, $L = \mathbb{Z}\cdot 1 + \mathbb{Z}\cdot z$ also. Since $z \in \mathcal{D}$, $|z| > 1$ and the same reasoning applies to the pair $1,z$. In other words, $|w_1| = 1$, $|w_2| = |z|$. Examination of the picture should convince the reader that this cannot occur unless already $z = w_2/w_1$ (i.e., $\gamma = \pm I$). (In more detail: the geometric observation shows that $w_1 = \pm 1$, $w_2 = \pm z$, whence $w_2/w_1 = \pm z$. But $w_2/w_1 = \gamma z \in P$, so this forces $w_2/w_1 = z$.) □

Finally, we remark that the identification of P with G/K endows P with a G-invariant measure (cf. 1.3), which is not the usual Lebesgue measure in the complex plane. Up to a scalar this has the form $\frac{1}{y^2}\,dx\,dy$, and $\bar{\mathcal{D}}$ turns out to have volume $\pi^2/3$ (see Gel'fand et al. for details). Since $\bar{\mathcal{D}}$ is noncompact this finiteness of volume is of special interest; this will be discussed below for the general case of $SL(n,R)$.

§10. Siegel sets in GL(n,R)

This section is based on Borel [5,§1], cf. also Borel, Harish-Chandra [1].

10.1 Iwasawa decomposition

We follow Bourbaki [3, Ch.7, §3, Example 7], where the following elementary lemma is proved.

LEMMA. *If* $(f_1,\dots,f_n)$ *is a basis of* $\mathbb{R}^n$, *then there exists a unique orthonormal basis* $(d_1,\dots,d_n)$ *of* $\mathbb{R}^n$ *such that* $f_i = \beta_{1i}d_1 + \dots + \beta_{ii}d_i$ $(i=1,\dots,n)$ *and all* $\beta_{ii} > 0$. *Moreover,* d_i *and* β_{ij} *depend continuously on* $f_1,\dots,f_n$.

Throughout this section, $G = GL(n,\mathbb{R})$, $K = O(n,\mathbb{R})$, $A = \{diag(a_1,\dots,a_n) \in G \mid a_i > 0 \text{ all } i\}$,

$$U = \left\{\begin{pmatrix} 1 & & * \\ & 1 & \\ & & \ddots & \\ 0 & & & 1\end{pmatrix} \in G\right\} = \text{upper triangular unipotent group.}$$

Clearly, K, A, U are closed subgroups of G. (The letter N is more commonly used than U, but we have another use in mind for N.) The following theorem is classical; its generalization to an arbitrary reductive Lie group is known as the Iwasawa decomposition.

THEOREM. *The natural product map* $K \times A \times U \to G$ *is a homeomorphism.*

Proof. Let $(e_1,\dots,e_n)$ be the canonical basis of $\mathbb{R}^n$. If $g \in GL(n,\mathbb{R})$ is arbitrary, write $f_i = g\cdot e_i$. Then use the above lemma to find an orthonormal basis $(d_1,\dots,d_n)$ for which $f_i = \beta_{1i}d_1 + \dots + \beta_{ii}d_i$ and all $\beta_{ii} > 0$. Let $k \in K$ be the orthogonal matrix sending $e_i \mapsto d_i$ (this specifies k uniquely), so that $k^{-1} f_i = \beta_{1i}e_1 + \dots + \beta_{ii}e_i$. Let $b = (\beta_{ij})$. Evidently b belongs to the upper triangular subgroup equal to the semidirect product A.U (homeomorphic to $A \times U$ under the obvious map). So write $b = a.n$. Finally, $g = kb = kan$. Uniqueness and bicontinuity of the product decomposition now follow from the lemma and the remark just made about A.U. □

Notation. If $g \in G$, we may write $g = k_g a_g n_g$ for the Iwasawa decomposition of g. When referring to the coordinates of the A-part, we write a_i instead of the more cumbersome a_{ii}, when convenient.

10.2 Siegel sets

We now use the Iwasawa decomposition $G = K.A.U$ to define certain subsets of G which (for suitable parameters) will be approximations to fundamental domains for $\Gamma = GL(n,\mathbb{Z})$ in G.

If t, u are positive real numbers, let

$$\Sigma_{t,u} = K.A_t.U_u ,$$

where $A_t = \{a \in A \mid a_i/a_{i+1} \leq t, \quad i = 1,\dots,n-1\}$

and $U_u = \{n \in U \mid |n_{ij}| \leq u \text{ for } i < j\}$.

It is obvious that K and U_u are compact sets, whereas A_t is not, so $\Sigma_{t,u}$ is noncompact (which is what we would expect, on the basis of the example in §9). To anticipate §11, the Iwasawa decomposition and construction of Siegel sets adapt immediately to $G^* = SL(n,\mathbb{R})$, $K^* = SO(n,\mathbb{R})$, $A^* = A \cap G^*$, $U^* = U \cap G^* = U$. So we can ask how $\Sigma^*_{t,u} = \Sigma_{t,u} \cap G^*$ will compare with the fundamental domain $\mathcal{D}$ for $\Gamma^* = SL(2,\mathbb{Z})$ in $\mathcal{P}$ which we already discussed in the case $n = 2$.

The notational convention of Borel forces us to reverse left and right, however. We described $\mathcal{P}$ as G^*/K^* ($G^* = SL(2,\mathbb{R})$ now), but instead we want to use $K^*\backslash G^*$ and let Γ^* act on the right. This is easily done by applying to G^* the involutive antiautomorphism $\begin{pmatrix} a & b \\ c & d \end{pmatrix} \mapsto \begin{pmatrix} 0 & 1 \\ 1 & 0 \end{pmatrix} {}^t\!\begin{pmatrix} a & b \\ c & d \end{pmatrix} \begin{pmatrix} 0 & 1 \\ 1 & 0 \end{pmatrix} = \begin{pmatrix} d & b \\ c & a \end{pmatrix}$. Now G^* acts on $\mathcal{P}$ on the right by $z \mapsto \dfrac{dz+b}{cz+a}$. The reader should check that the isotropy group of i is still K^*, so $\mathcal{P}$ identifies with $K^*\backslash G^*$. (Because of the Iwasawa decomposition, this looks like A^*U^*.) The reader should also verify that $\mathcal{D}$ is still a fundamental domain for Γ^* in $\mathcal{P}$ (Γ^* acting in the new way).

Now we just compute $i.\Sigma^*_{t,u} = i.K^*A^*_tU^*_u = i.A^*_tU^*_u$. A typical element of $A^*_tU^*_u$ is $\begin{pmatrix} a & 0 \\ 0 & a^{-1} \end{pmatrix}\begin{pmatrix} 1 & n \\ 0 & 1 \end{pmatrix} = \begin{pmatrix} a & an \\ 0 & a^{-1} \end{pmatrix}$ with $\dfrac{a}{a^{-1}} = a^2 \leq t$, $|n| \leq u$. Therefore $i.A^*_tU^*_u$ consists of all $z \in \mathcal{P}$ with $\operatorname{Im} z \geq \frac{1}{t}$ and $|\operatorname{Re} z| \leq u$ (check!). We conclude that $i.\Sigma^*_{t,u}$ contains

$\mathcal{D}$ iff $t \geq 2/\sqrt{3}$, $u \geq 1/2$ (and these are the best possible bounds). The reader should draw a picture and observe that $\Sigma^*_{t,u}$ cannot be a fundamental domain in the strict sense of §9. The problem is not just at the boundary; rather, $\Sigma^*_{t,u}$ (for $t \geq 2/\sqrt{3}$, $u \geq 1/2$) will meet some of its Γ^*-translates in the interior. In spite of this, we will be able (later on) to prove that such a Siegel set intersects only finitely many of its Γ^*-translates.

10.3 Minimum principle

We consider again the general case $G = GL(n,\mathbb{R})$, $\Gamma = GL(n,\mathbb{Z}) = \{g \in M(n,\mathbb{Z}) \mid \det g = \pm 1\}$. Geometrically, it is the space $K \backslash G$ which interests us (for reasons explained in §11), but it is more convenient to work directly in G. The reader should be aware, however, that it would be equivalent to study the (right) action of Γ on either G or $K \backslash G$; from the viewpoint of fundamental domains, the compact set K causes no trouble (which is reflected in the definition of Siegel sets).

Our main goal is to prove that for suitable t, u, the Γ-translates of $\Sigma_{t,u}$ will cover G; this is roughly the first requirement for $\Sigma_{t,u}$ to be a fundamental set. (The fact that $\Sigma_{t,u}$ intersects only finitely many of its Γ-translates is more subtle, so the proof will require more delicate properties of G.) In §9 the corresponding statement was proved by constructing certain vectors of minimal length in a lattice. Here we use a similar method.

Let $(e_1,\ldots,e_n)$ be the canonical basis of $\mathbb{R}^n$. To avoid confusion with absolute value in $\mathbb{R}$, denote euclidean length by $\|x\|$. Define $\Phi : G \to \mathbb{R}^{>0}$ by $\Phi(g) = \|g.e_1\|$ $(=\sqrt{g_{11}^2+g_{21}^2+\ldots g_{n1}^2})$. Φ is obviously continuous.

Remarks (a) If $g = k.a.n$ (Iwasawa decomposition), then the facts that k is orthogonal and $n.e_1 = e_1$ imply

$\Phi(g) = \|k.a.n.e_1\| = \|a.n.e_1\| = \|a.e_1\| = a_1 = \Phi(a)$.

(b) If $g \in G$, $g\Gamma.e_1 \subset g(\mathbb{Z}e_1 + \ldots + \mathbb{Z}e_n - \{0\})$ = nonzero points of a lattice in $\mathbb{R}^n$, so Φ achieves a positive minimum on the coset $g\Gamma$.

In view of (b) the following theorem makes sense, and has as an immediate corollary the statement that $G = \Sigma_{t,u}\cdot\Gamma$ whenever $t \geq 2/\sqrt{3}$, $u \geq 1/2$ (compare these values of t, u with those found in the discussion of $G^* = SL(2,\mathbb{R})$ in 10.2).

THEOREM. *If $g \in G$, the minimum of Φ on $g\Gamma$ is achieved at a point of $\Sigma_0 = \Sigma_{2/\sqrt{3},1/2}$.*

Proof. This proceeds in several steps. In view of Remark (a), the A-component is most crucial.

(1) $U = U_{1/2}(U \cap \Gamma)$: Let $u = (u_{ij}) \in U$, $z = (z_{ij}) \in U \cap \Gamma$. Then for $i < j$, $(uz)_{ij} = z_{ij} + u_{i,i+1}z_{i+1,j} + \ldots + u_{ij}$. An obvious induction on pairs (i,j) with $i < j$, starting at $(n-1,n)$, allows us to solve for integers z_{ij} such that finally $|(uz)_{ij}| \leq 1/2$ $(i<j)$. At the first step we just have to assure $|z_{ij} + u_{ij}| \leq 1/2$. (The rest is left to the reader as an exercise).

(2) In view of (1), *if Φ takes its minimum on a coset at $g = k.a.n$, then we may assume* $n \in U_{1/2}$ (use Remark (a) to see that altering the U-component does not affect Φ).

(3) *If Φ takes its minimum on a coset at* $g = k.a.n$, *then* $a_1/a_2 \leq 2/\sqrt{3}$. Using (2) we may assume $|n_{ij}| \leq 1/2$ for $i < j$.

Let $z = \left(\begin{array}{c|c} \begin{matrix} 0 & 1 \\ 1 & 0 \end{matrix} & 0 \\ \hline 0 & I \end{array}\right)$, so $\det z = -1$ and $z \in \Gamma$. Then $(gz).e_1 = g.e_2 = k.a.n.e_2 = k.a.(e_2 + n_{12}e_1) = k(a_2.e_2 + a_1n_{12}.e_1)$, hence $\Phi(gz)^2 = a_1^2n_{12}^2 + a_2^2 \leq a_1^2/4 + a_2^2$ (since $|n_{12}| \leq 1/2$). Minimality forces $\Phi(g)^2 = a_1^2 \leq a_1^2/4 + a_2^2$, so $a_1/a_2 \leq 2/\sqrt{3}$ as required.

(4) Now we can prove the theorem, using induction on n. Our previous set-up does not include the case $n = 1$ unless we make the

convention $A = A_t$ (all t). So $\Sigma_0 = G$ and there is nothing to prove. (Exercise. Check the argument below for passage from $n = 1$ to $n = 2$.)

Now let Φ take its minimum on a coset $g\Gamma$ at $g = k.a.n.$ Write $k^{-1}g = \begin{pmatrix} a_1 & * \\ 0 & b \end{pmatrix}$, with $b \in GL(n-1,\mathbb{R})$. By induction there exists $z' \in GL(n-1,\mathbb{Z})$ such that $bz' \in \Sigma_0^{(n-1)}$. Write $bz' = k'a'n'$, and let $z = \begin{pmatrix} 1 & 0 \\ 0 & z' \end{pmatrix} \in \Gamma$. Then
$k^{-1}gz = \begin{pmatrix} a_1 & * \\ 0 & k'a'n' \end{pmatrix} = k''a''n''$, where $k'' = k\begin{pmatrix} 1 & 0 \\ 0 & k' \end{pmatrix} \in K$,
$a'' = \begin{pmatrix} a_1 & 0 \\ 0 & a' \end{pmatrix} \in A$, $n'' = \begin{pmatrix} 1 & * \\ 0 & n' \end{pmatrix} \in U$. But by induction, $a''_i/a''_{i+1} \leq 2/\sqrt{3}$ $(i=2,\ldots,n)$. Since $ze_1 = e_1$, $\Phi(gz) = \Phi(g)$ is still the minimum value of Φ on the coset $g\Gamma$. According to (3), $a''_1/a''_2 \leq 2/\sqrt{3}$. Therefore $gz \in KA_{2/\sqrt{3}}U$, and if we apply (2) we finally get $gz^* \in KA_{2/\sqrt{3}}U_{1/2}$ (for some $z^* \in \Gamma$) $= \Sigma_0$.□

Exercise. For $n \geq 2$, show that the bounds $2/\sqrt{3}, 1/2$ are best possible. (For $n = 2$ this follows from §9 and 10.2, if we adapt to SL.)

§11. Applications

We continue to follow Borel [5, §1-2].

11.1 Siegel sets in $SL(n,\mathbb{R})$

In this section $G^* = SL(n,\mathbb{R})$, $K^* = SO(n,\mathbb{R})$, $A^* = A \cap SL(n,\mathbb{R})$, $U^* = U \cap G^*$ $(= U)$, $B^* = A^*U^*$ ($\subset$ upper triangular subgroup of G^*), $\Sigma^*_{t,u} = \Sigma_{t,u} \cap G^* = K^*A^*_tU^*_u$. It is trivial to verify the Iwasawa decomposition theorem for $G^* = K^*A^*U^*$. Moreover, if $\Gamma^* = SL(n,\mathbb{Z})$, then the theorem of 10.3 obviously carries over to G^*. The main difference between G and G^* is brought out in the following.

THEOREM. $\Sigma^*_{t,u}$ has finite Haar measure (for any t,u).

Exercise. Show that in G, $\Sigma_{t,u}$ does not have finite measure. [Use properties of the various Haar measures discussed below.]

We know in general that a compact subset of a locally compact group has finite measure; this will suffice for K^* and U_u^*. But for A^* we have to do some actual calculation; this is best done by realizing Haar measure in some concrete way. Moreover, we must compare Haar measure dg on G^* with Haar measures dk, da, dn on K^*, A^*, U^*.

Since K^* is compact and A^* abelian, each of these groups is unimodular (1.2). So is U^* (essentially because it is a nilpotent Lie group): this may be seen by comparing U^* with the locally compact (additive) group $\mathbb{R}^m$ $(m = n(n-1)/2)$ to which U^* is homeomorphic via $(n_{ij}) \overset{\phi}{\longmapsto} (\ldots, n_{ij}, \ldots)_{i<j}$.

Exercise. Φ transfers Lebesgue measure on $\mathbb{R}^m$ to left/right Haar measure on U^* (so U^* is unimodular, in particular). (This exercise is rather nontrivial, but the reader is urged to attempt it before looking at the proof in Bourbaki [3,Ch.7, §3, no. 3, Ex.3].)

Now A^* acts (by inner automorphisms in G^*) on U^*, so the module of $\mathrm{Int}(a)|_{U^*}$ is defined for each $a \in A^*$. But $\mathrm{Int}(a)$ simply multiplies the (i,j) entry of $n \in U^*$ by a_i/a_j (check!); for $i < j$ this multiplies the corresponding coordinate of $\mathbb{R}^m$ by a_i/a_j, thereby changing Lebesgue measure by that factor in each component of $\mathbb{R}^m$. Using the exercise above, we see that $\mathrm{Int}(a)|_{U^*}$ has module $\rho(a) = \prod_{i<j} a_i/a_j$.

But B^* is (topologically and algebraically) the semidirect product of A^* and U^* (U^* normal), so an easy calculation shows that left and right Haar measure on B^* differ by the factor ρ , cf. Bourbaki [3, Ch.7, §2, no. 9, Prop. 14]. The product measure $da.dn$ is an obvious choice for left Haar measure, so right Haar measure must be (up to a scalar multiple) $\rho(a)\, da.dn$. Recall what this symbolism means: to calculate right Haar measure of a set of the form $A_o^* U_o^*$ in B^*, we have to evaluate

$$\int_{A_o^* \times U_o^*} \rho(a)\, da\, dn \quad .$$

Now let $\pi : K^* \times A^* \times U^* \to G^*$ be the product map (which is a homeomorphism). We claim that π^{-1} transfers dg to (a scalar multiple of) the product measure $\rho(a)\, dk\, da\, dn$. To see this, compare $\pi' : K^* \times B^* \to G^*$. It is clear that $(\pi')^{-1}$ transfers dg to a measure left-invariant under K^* and right-invariant under B^*, hence to a measure of the form $dk\, d_r b$ where $d_r b$ is a right Haar measure on B^*. So the underlined assertion follows from the previous remarks.

This shows that $\int_{\Sigma^*_{t,u}} dg$ is equal to $c \int_{A_t^*} \rho(a)\, da$, for some positive c, in view of the fact that K^*, U_u^* have finite measure. To evaluate the integral over A_t^* we must realize da more concretely, as follows. Let $\beta : A^* \to (\mathbb{R}^{>0})^{n-1}$ be defined by $\beta(a) = (\beta_1(a), \ldots, \beta_{n-1}(a)) = (a_1/a_2, a_2/a_3, \ldots, a_{n-1}/a_n)$. Since $\det a = \prod_{i=1}^{n} a_i = 1$, knowledge of $\beta_1(a), \ldots, \beta_{n-1}(a)$ allows us to recover a in a unique manner. Therefore it is immediate that β is an isomorphism of topological groups.

On the other hand, let $\mathfrak{a}^*$ = set of all matrices diag $(z_1, \ldots, z_n)$ of trace 0. (This is in fact the Lie algebra of the connected Lie group A^*.) Define $\gamma : \mathfrak{a}^* \to \mathbb{R}^{n-1}$ by $\gamma(z) = (\gamma_1(z), \ldots, \gamma_{n-1}(z)) = (z_1 - z_2, z_2 - z_3, \ldots, z_{n-1} - z_n)$. Since $\sum_{i=1}^{n} z_i = 0$, it is clear that γ is an isomorphism of (additive) topological groups, $\mathbb{R}^{n-1}$ having the usual topology.

We can define in the obvious way homomorphisms $\log : A^* \to \mathfrak{a}^*$ and $\exp : \mathfrak{a}^* \to A^*$, which are inverse to each other, and similarly maps between $(\mathbb{R}^{>0})^{n-1}$ and $\mathbb{R}^{n-1}$. The diagram then commutes:

$$\begin{array}{ccc} A^* & \xrightarrow{\beta} & (\mathbb{R}^{>0})^{n-1} \\ \log\downarrow\uparrow\exp & & \log\downarrow\uparrow\exp \\ \mathfrak{a}^* & \xrightarrow{\gamma} & \mathbb{R}^{n-1} \end{array}$$

With this set-up we can replace any one of these locally compact groups by any other; so the Haar measure da on A^* might as well be taken to correspond to Haar measure (= Lebesgue measure) on $\mathbb{R}^{n-1}$.

What does the function $\rho(a)$ look like on $\mathbb{R}^{n-1}$? Notice that $\rho(a) = \prod_{i=1}^{n-1} \beta_i(a)^{r_i}$ for certain positive integers r_i, because $(i < j)$ $a_i/a_j = a_i/a_{i+1} \cdot a_{i+1}/a_{i+2} \cdots$ So $\rho(a)$ becomes $\prod_{i=1}^{n-1}$ (exp $r_i\gamma_i$ (log a)). Finally,

$$\int_{A_t^*} \rho(a)\, da = \prod_{i=1}^{n-1} \left(\int_{-\infty}^{\log t} (\exp r_i y_i)\, dy_i \right) < \infty$$

(because $r_i \geq 0$). Here $\prod dy_i$ is standard measure on $\mathbb{R}^{n-1}$. □

<u>Exercise</u>. For $SL(2,\mathbb{R})$, normalize the choice of Haar measure and compute explicitly the measure of $\Sigma^*_{2/\sqrt{3},1/2}$.

<u>Exercise</u>. With G, G^* as before, show that both G and G^* are unimodular. (This is a special case of the fact that any reductive Lie group is unimodular.)

11.2 Reduction of positive definite quadratic forms

A quadratic form on $\mathbb{R}^n$ is a function $F : \mathbb{R}^n \to \mathbb{R}$ obtained from a symmetric bilinear form $F'(x,y)$ by the rule $F(x) = F'(x,x)$. Given F, the form can be recovered uniquely, since $F(x+y) - F(x) - F(y) = 2F'(x,y)$. F may be identified with the (symmetric) $n \times n$ matrix $(F'(e_i,e_j))$, where $(e_1,\ldots,e_n)$ is the canonical basis of $\mathbb{R}^n$.

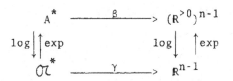

Call F <u>positive definite</u> if $F(x) \geq 0$ for all x and $F(x) = 0$ iff $x = 0$. (For example, the usual euclidean length, with matrix I.) It is well known that F is positive definite iff all diagonal minors of (the matrix of) F are positive. $G = GL(n,\mathbb{R})$ acts (on the right) on the set H of all positive definite forms, viewed as matrices, by the rule $F \mapsto F[g] = {}^t gFg$. This just amounts to replacing the canonical basis by a new basis $(ge_1,\ldots,ge_n)$; therefore $F[g]$ is again positive definite. Concretely, we have:

$$F[g]_{ij} = ({}^t gFg)_{ij} = \sum_{k,\ell} g_{ki} F(e_k,e_\ell) g_{\ell j} = F(ge_i,ge_j).$$

<u>Exercise</u>. G acts transitively on H.

The isotropy group of I is $\{g \in G \mid {}^t gIg = I\} = \{g \in G \mid {}^t gg = I\}$ $= O(n,\mathbb{R}) = K$. Therefore the set H may be identified with $K \backslash G$ and topologized accordingly. We may ask now how to choose reasonable representatives of the Γ-orbits in H (this is the problem of integral equivalence of positive definite forms). The projection map from G to H sends a Siegel set $\Sigma_{t,u}$ to

$\Sigma'_{t^2,u} = \{{}^t n.a.n \mid a \in A_{t^2},\ n \in N_u\}$, since for $g = k.a.n$, $I[g] = {}^t na^2 n$. Theorem 10.3 now becomes a classical result about reduction:

<u>THEOREM</u>. $H = \Sigma'_{t,u}[\Gamma]$ <u>whenever</u> $t \geq 4/3$, $u \geq 1/2$.

Similar considerations apply to forms with matrix of det 1, on which $SL(n,\mathbb{R})$ acts. Here we get the additional result that $H^{(1)}/SL(n,\mathbb{Z})$ has finite invariant measure ($H^{(1)}$ the space of forms in question).

§12. <u>BN-pairs</u>

In this section we consider more carefully the algebraic structure of GL_n, in preparation for the proof of the "Siegel property" in §13. For a concrete description of what happens in GL_n, see Borel [5, §3]. Here we use instead the axiomatic framework of BN-pairs (now known as Tits systems) invented by Tits [1]. This same formalism applies equally well to other reductive Lie groups which arise in connection with arithmetic groups, as well as to p-adic groups (cf. §15 below for an example). References for BN-pairs include Bourbaki [2, Chapter 4], Carter [1], Humphreys [1,§29], Richen [1].

12.1 Axioms and Bruhat decomposition

A BN-pair is a 4-tuple (G,B,N,R), where G is a group generated by subgroups B and N, $H = B \cap N$ is normal in N, R is a finite set of involutions which generate the Weyl group $W = N/H$, and the following axioms hold:

(BN1) If $r \in R$, $w \in W$, then $rBw \subset BwB \cup BrwB$.

(BN2) If $r \in R$, $rBr \neq B$.

Remarks. (a) Notations such as wB, Bw, BwB $(w \in W)$ are allowed, because two representatives of w in N differ by something in $H \subset B$.

(b) The Weyl group W may be either finite or infinite; later we will require it to be finite.

(c) Axiom (BN1) expresses the fact that the product of the two double cosets BrB and BwB (with $r \in R$) is included in the union of two double cosets BwB and $BrwB$. An even sharper statement will be deduced below.

(d) By taking inverses, and using the fact that $r^{-1} = r$, (BN1) becomes:

If $r \in R$, $w \in W$, then $wBr \subset BwB \cup BwrB$.

(So the asymmetry is only apparent).

Exercise. (Tits, Sém. Bourbaki, No. 288). If (BN2) is replaced by "$rBr^{-1} \not\subset B$ $(r \in R)$," then the assumption that R consists of involutions can be deduced from the other assumptions on (G,B,N,R).

Example. k = field

$G = GL(n,k)$

B = upper triangular matrices

N = monomial matrices (having exactly one nonzero entry in each row and column)

(So H is the diagonal group, clearly normal in N, and W = N/H is isomorphic to the symmetric group S_n.)

$R = \{(i, i+1), 1 \leq i \leq n-1\}$

The BN-axioms, except for (BN1), are easy to verify.
For (BN1), see for example Bourbaki [2, Ch.4, §2, no. 2]. Note that W is finite here.

Remark. Card(R) is called the rank of the BN-pair. For the example just discussed the rank is n-1. In Lie theory GL_n is viewed as having rank n, because the diagonal group H in $GL(n,\mathbb{R})$ has dimension n; however, the one-dimensional subgroup of scalars plays no role in the BN-structure.

Exercise. Construct a BN-pair for SL(n,k) analogous to that for GL(n,k). What is the rank?

For $J \subset R$, let W_J = subgroup of W generated by J. Set $G_J = BW_JB$.

THEOREM. (a) *G_J is a subgroup of G, for $J \subset R$. (In particular, $G_R = BWB = G$ since G is generated by B,N; this is the Bruhat decomposition of G.)*
(b) *If $BwB = Bw'B$, then $w = w'$.*

Proof (doesn't require axiom (BN2)).
(a) G_J is inverse-closed, so it suffices to show that G_J is closed under left multiplication by B, J (the former being obvious). If $r \in J$, then thanks to (BN1), $rBW_JB \subset BW_JB \cup BrW_JB \subset G_J$.
(b) Here we use induction on length: $\ell(w)$ is the smallest possible k such that $w = r_1 \ldots r_k (r_i \in R)$, and then the expression $w = r_1 \ldots r_k$ is called reduced. ($\ell(1) = 0$, $\ell(r) = 1$ for $r \in R$, $\ell(w) > 1$ otherwise.) We may assume without loss of

generality that $\ell(w) \leq \ell(w')$ and use induction on $\ell(w)$. If $\ell(w) = 0$, $w = 1$ and therefore $BwB = Bw'B$ implies $B = Bw'B$. So a representative of w' in N lies in B, forcing (since $B \cap N = H$, $W = N/H$) $w' = 1 = w$. For the induction step, assume $BwB = Bw'B$ with $\ell(w) \geq 1$. Write $w = rw^*$ $(r \in R)$ with $\ell(w^*) = \ell(w)-1$. Now $BwB = Bw'B$ implies $rw^*B \subset Bw'B$. Since $r^2 = 1$, $w^*B \subset rBw'B \subset Bw'B \cup Brw'B$. Two cases arise:

(i) $w^* \in Bw'B \Rightarrow Bw^*B = Bw'B \Rightarrow w^* = w'$

(by induction), contradicting $\ell(w^*) < \ell(w) \leq \ell(w')$.

(ii) $w^* \in Brw'B \Rightarrow Bw^*B = Brw'B \Rightarrow w^* = rw'$

(by induction, since $\ell(w^*) \leq \ell(rw')$ is clear)

$\Rightarrow rw^* = w'$ $(r^2=1)$ or $w = w'$, as desired. □

12.2 Parabolic subgroups

Here we get some precise information about the inclusion expressed in (BN1), related to length of elements of W, and use this information to show that the only subgroups of G containing B are the groups G_J of Theorem 1. In view of Theorem 1(b) and the fact that $r \in R$ has order 2, it follows from (BN1) that $rBw \subset BrwB$ iff $rBw \cap BwB = \emptyset$.

LEMMA 1. (a) $\ell(rw) \geq \ell(w) \Rightarrow rBw \subset BrwB$.

(b) $\ell(rw) \leq \ell(w) \Rightarrow rBw \cap BwB \neq \emptyset$.

(c) $\ell(rw) = \ell(w) \pm 1$ $(r \in R,\ w \in W)$.

Proof. By the remark preceding the lemma, (c) follows at once from (a), (b).

(a) Use induction on $\ell(w)$, $\ell(w) = 0$ being obvious. Write $w = w'r'$, $\ell(w') = \ell(w) - 1$, $r' \in R$. Suppose the conclusion false, i.e., $rBw \cap BwB \neq \emptyset$. Multiplying on the right by r' gives

(*) $$rBw' \cap BwBr' \neq \emptyset .$$

But $\ell(rw') \geq \ell(rw'r')-1 = \ell(rw)-1 \geq \ell(w)-1 = \ell(w')$. By induction,

$rBw' \subset Brw'B$. By (*), this implies $BwBr' \cap Brw'B \neq \emptyset$. But (using the symmetric form of (BN1)) $BwBr' \subset BwB \cup Bwr'B = BwB \cup Bw'B$. Therefore $Brw'B$ intersects (hence equals) one of these double cosets. By Theorem 1(b), either $w = rw'$ (absurd, since $\ell(w') < \ell(w) \leq \ell(rw)$ by assumption) or $w' = rw'$ (absurd since $r \neq 1$ by assumption). Contradiction.

(b) Here we apply (a), along with both (BN1) and (BN2):

$$\text{(BN1)} \quad rBr \subset BrB \cup B$$

$$\text{(BN2)} \quad rBr \cap Brb \neq \emptyset \quad .$$

Multiplying by rw on the right, $rBw \cap BrBrw \neq \emptyset$. But $\ell(r^2w) = \ell(w) \geq \ell(rw)$ by hypothesis, so part (a) says $BrBrw \subset BwB$, whence $rBw \cap BwB \neq \emptyset$. □

Exercise. If $w = r_1 \ldots r_k$ (reduced), then $BwB = (Br_1B)(Br_2B) \ldots (Br_kB)$.

LEMMA 2. *Let $w = r_1 \ldots r_k$ (reduced), $J = \{r_1, \ldots, r_k\}$. Then $\langle B, wBw^{-1} \rangle = \langle B, w \rangle = BW_JB$.*

Proof. Write $r_1w = w'$, so $\ell(w') < \ell(w)$. Lemma 1(b) says that $r_1Bw \cap BwB \neq \emptyset$, forcing $r_1 \in BwBw^{-1}B \subset \langle B, wBw^{-1} \rangle$. Proceeding by induction on length, $\{r_2, \ldots, r_k\} \subset \langle B, w'Bw'^{-1} \rangle = \langle B, r_1wBw^{-1}r_1 \rangle \subset \langle B, wBw^{-1}, r_1 \rangle = \langle B, wBw^{-1} \rangle$. Therefore, $G_J = BW_JB \subset \langle B, wBw^{-1} \rangle \subset \langle B, w \rangle \subset G_J$ (Theorem 1 (a)), so equality holds throughout. □

Exercises. (a) $wBw^{-1} = B \Rightarrow w = 1$ (cf. (BN2)).

(b) The set of $r \in R$ in a reduced expression for $w \in W$ is determined uniquely by w.

(3) R is uniquely determined by (G, B, N) as the set of those $w \in W$ for which $B \cup BwB$ is a group. [Use Theorem 1(a) & Lemma 2.]

LEMMA 3. *R is a minimal generating set for W.*

Proof. Suppose not, so some $R' = R - \{r\}$ generates W. The axioms obviously remain valid for (G,B,N,R'). Let $r = r_1 \ldots r_k$ $(r_i \in R')$ be a reduced expression (so $k > 1$). Then Lemma 2, applied to (G,B,N,R'), shows that $\{r_1,\ldots,r_k\} \subset B \cup BrB = \langle r,B\rangle$ (which is still a group!); since $k > 1$, this contradicts Theorem 1(b), applied to (G,B,N,R). □

LEMMA 4. Let $w \in W$, $J,K \subset R$. Then $wG_Jw^{-1} \subset G_K$ implies that $w \in G_K$.

Proof. B, $wBw^{-1} \subset G_K$, so by Lemma 2, $w \in \langle B,wBw^{-1}\rangle \subset G_K$. □

Call a subgroup of G parabolic if it contains a conjugate of B. We can now determine all those parabolic subgroups containing B.

THEOREM 2. (a) If M is a subgroup of G containing B, then $M = G_J$ for some $J \subset R$.

(b) If G_J is conjugate to G_K, $G_J = G_K$.

(c) $N_G(G_J) = G_J$ (in particular, $N_G(B) = B$).

(d) $W_J \subset W_K$ implies $J \subset K$.

(e) $G_J \subset G_K$ implies $J \subset K$.

(It follows at once from (e) that the lattice of subgroups of G containing B is isomorphic to the lattice of subsets of R.)

Proof. (a) If $M \supset B$, Theorem 1(a) implies that M is the union of certain BwB, so $M \subset G_J$ (J = union of the sets J_w of $r \in R$ appearing in reduced expressions for such w). But Lemma 2 forces $G_{J_w} \subset M$, whence $G_J \subset M$.

(b) Use Lemma 4; since $B \subset G_J$, G_K, these subgroups of $G = \langle B,N\rangle$ are conjugate iff they are conjugate by some $w \in W$.

(c) Again use Lemma 4.

(d) If $W_J \subset W_K$, each $r \in J$ is a product of elements of K, so (by Lemma 3) $r \in K$.

(e) If $G_J \subset G_K$, then $W_J \subset G_K$, forcing $W_J \subset W_K$ (since $B \cap N = H$).

In turn (part(d)) $J \subset K$. ∎

Example. When $G = GL(n,k)$, the subgroups containing B may be described concretely as stabilizers of "flags" $0 \subset [e_1,\ldots,e_{i_1}] \subset [e_1,\ldots,e_{i_2}] \subset \ldots \subset [e_1,\ldots,e_n]$, where $i_1 < i_2 < \ldots$ and where $(e_1,\ldots,e_n)$ is the standard basis of k^n. B itself is the stabilizer of the flag $0 \subset [e_1] \subset [e_1,e_2] \subset [e_1,e_2,e_3] \ldots \subset [e_1,\ldots,e_n] = k^n$.

If k is a finite field, then U(= upper triangular unipotent group) is readily seen to be a Sylow p-subgroup of G (p = char k), with $B = N_G(U)$. Then Theorem 2 (c) expresses a well known property of Sylow normalizers and groups including such. (Question: What happens if $n = 2$ and $k = \mathbf{F}_2$?)

12.3 Conjugates of B by W

Write $B^w = w^{-1}Bw$.

DEFINITION. The BN-pair (G,B,N,R) is saturated if $\bigcap_{n\in N} B^n$ $(= \bigcap_{w\in W} B^w) = H$. (In general, of course, this intersection includes H).

LEMMA 5. Let $N' = \langle H',N\rangle$, $H' = \bigcap_{n\in N} B^n$. Then N'/H' is canonically isomorphic to $N/H = W$, and if R' corresponds to R, then (G,B,N',R') is a saturated BN-pair with Weyl group $W' \cong W$.

Proof. H' is normalized by N and H', so H' is normal in $N' = NH'$. Moreover, $B \cap N'$ is N'-invariant, hence lies in H'; $H' \subset B \cap N'$ is clear, so $H' = B \cap N'$. Therefore $W' = N'/H' = NH'/H' \cong N/N \cap H' = N/H = W$, and the remaining BN-axioms are obvious ((BN1) and (BN2) being automatic). ∎

We assume henceforth (with little loss of generality) that

(G,B,N,R) <u>is saturated</u>. This holds for the BN-pair in GL(n,k) defined earlier, as well as for other interesting examples.

<u>LEMMA 6</u>. (a) <u>If</u> $\ell(rw) > \ell(w)$, $rBr \cap BwBw^{-1} \subset B$.

(b) <u>If</u> $\ell(wr) < \ell(w)$, $B^r \cap B^{wr} \subset B$.

<u>Proof</u>. (a) $rBr \subset B \cup BrB$ (BN1), so if (a) fails we must have $BrB \cap BwBw^{-1} \neq \emptyset$, or $BrBw \cap BwB \neq \emptyset$. But Lemma 1(a) applies, so $BrBw \subset BrwB$; therefore $BrwB \cap BwB \neq \emptyset$, or $BrwB = BwB$. By Theorem 1 (b), $rw = w$, so $r = 1$, contradicting the fact that r has order 2 (or (BN2)).

(b) Since $R = R^{-1}$, $\ell(w) = \ell(w^{-1})$. Now $\ell(wr) < \ell(w)$ implies $\ell(rw^{-1}) < \ell(rrw^{-1})$, so $B^r \cap B^{wr} \subset B$, thanks to (a) . □

<u>LEMMA 7</u>. <u>If</u> $\ell(rw) > \ell(w)$, $B \cap B^{rw} \subset B \cap B^w$.

<u>Proof</u>. It suffices to show that $B \cap B^{rw} \subset B^w$. Using Lemma 6(a) we have $(B \cap B^{rw})^{w^{-1}} = B^{w^{-1}} \cap B^r \subset B$. □

From now on we <u>assume that</u> W <u>is finite</u>. We want to locate in W an element w_0 for which $B \cap B^{w_0} = H$ (which requires, of course, that the BN-pair at least be saturated). For $G = GL(n,k)$, the permutation sending $i \mapsto n + 1 - i$ clearly does the trick: B^{w_0} is then the lower triangular group. By proving that W is a (finite) "Coxeter group" and studying W in a geometric setting as a group generated by reflections (in a real euclidean space), one can prove that there exists one (and only one) w_0 of this type. Rather than take this route, we simply add a new axiom, which the reader may verify directly for the Weyl group S_n of GL(n,k). Call $w \in W$ <u>left maximal</u> if $\ell(rw) < \ell(w)$ for all $r \in R$. (Since W is finite, such elements clearly exist.) We assume now:

(BN3) W <u>has a unique left maximal element</u> w_0.

<u>Exercise</u>. Verify (BN3) for S_n. [Suggestion: In S_n the length of an element σ relative to the generators (1,2),

$(2,3),\ldots,(n-1,n)$ can be shown to equal the number of pairs i,j $(i < j)$ for which $\sigma(i) > \sigma(j)$. Then show that a left maximal σ must satisfy $\sigma(i) > \sigma(j)$ for <u>all</u> $i < j$.]

Since a "maximal" w (one for which $\ell(rw) < \ell(w)$ and $\ell(wr) < \ell(w)$ for all $r \in R$) is both left and right maximal, and since the inverse of a left maximal element is right maximal, (BN3) implies that w_0 <u>is the unique maximal element of</u> W <u>and that</u> $w_0 = w_0^{-1}$.

<u>LEMMA 8</u>. $\ell(w_0 w) = \ell(w_0) - \ell(w)$ <u>for all</u> $w \in W$.

<u>Proof</u>. Let $w = r_1 \ldots r_k$ (reduced), so $w^{-1} = r_k \ldots r_1$ is also a reduced expression. If w^{-1} is not already left maximal, get some $r_{k+1} \in R$ such that $\ell(r_{k+1} r_k \ldots r_1) = k+1$. Continue until $w_0 = r_\ell \ldots r_{k+1} r_k \ldots r_1$ with $\ell(w_0) = \ell$. Then $\ell(w_0 w) = \ell(r_\ell \ldots r_{k+1})$ is obviously equal to $\ell(w_0) - \ell(w)$ $(=\ell - k)$. □

<u>LEMMA 9</u>. $B \cap B^{w_0} = H$.

<u>Proof</u>. (Here we use saturation for the first time.) Let $w \in W$. As in the proof of Lemma 8, we may continue to the left until we get a left maximal element $r_1 \ldots r_s w$, increasing length by one at each step. Then $w_0 = r_1 \ldots r_s w$ by (BN3). Iteration of Lemma 7 shows $B \cap B^{w_0} \subset B \cap B^w$. Since the BN-pair is saturated, this shows: $H \subset B \cap B^{w_0} \subset \bigcap_{w \in W} B^w = H$. □

<u>LEMMA 10</u>. <u>If</u> $\ell(wr) > \ell(w)$, $B \cap B^{w_0 r} \subset B \cap B^w$.

<u>Proof</u>. Write $w_0 r w^{-1} = r_1 \ldots r_k$ (reduced), $w_j = r_j \ldots r_k w$ (so $w_1 = w_0 r$). By assumption $\ell(wr) = \ell(rw^{-1}) = \ell(w) + 1$ (see Lemma 1(c)), and by Lemma 8, $k = \ell(w_0 r w^{-1}) = \ell(w_0) - \ell(rw^{-1})$ $= \ell(w_0) - \ell(w) - 1$. Therefore, $\ell(w_0 r) = \ell(w_0) - 1$ (by maximality or Lemma 8) $= k + \ell(w)$. This shows that length <u>increases</u> (precisely by 1) at each step of the sequence $w, w_k, w_{k-1}, \ldots, w_1 = w_0 r$. So we may apply Lemma 7 repeatedly:

$$\ldots\ B \cap B^{w_k} \subset B \cap B^{w}$$
$$\ldots\ B \cap B^{w_j} \subset B \cap B^{w_j+1} \subset \ldots$$

Finally,
$$B \cap B^{w_0 r} = B \cap B^{w_1} \subset \ldots \subset B \cap B^{w}\ .\square$$

LEMMA 11. (a) If $\ell(wr) > \ell(w)$, $B = (B \cap B^r)(B \cap B^w)$.

(b) $B = (B \cap B^r)(B \cap B^{w_0 r})$.

(Note that the order of the two factors can be reversed by taking inverses. Also, the two factors in (b) clearly intersect just in H, so the decomposition is "almost" unique.)

Proof. (a) If $\ell(wr) > \ell(w)$, $\ell(rw^{-1}) > \ell(w^{-1})$. By Lemma 1(a), $rBw^{-1} \subset Brw^{-1}B$, or $B \subset (rBr)(w^{-1}Bw)$. By Lemma 6(a) $Bw^{-1}Bw \cap B^r \subset B$, so the factors of B in the product $(rBr)(w^{-1}Bw)$ are themselves in B.
(b) w_0 being maximal, $\ell(w_0rr) > \ell(w_0r)$, so (a) implies that $B = (B \cap B^r)(B \cap B^{w_0 r})$. $\square$

Write $B_w = B \cap B^w$, $B_w^- = B \cap B^{w_0 w}$. Lemma 11(b) thus asserts that $B = B_r B_r^-$ $(r \in R)$. Our aim is to extend this to arbitrary $w \in W$.

LEMMA 12. (a) If $\ell(wr) > \ell(w)$, $B_{wr}^- = (B_r^-)(B_w^-)^r$.

(b) If $\ell(w) > \ell(wr)$, $B_w^- = (B_r^-)(B_{wr}^-)^r$.

(In each case the two factors intersect just in H.)

Proof. (b) follows from (a), using wr in place of w. For (a), $\ell(w_0wrr) = \ell(w_0w) = \ell(w_0) - \ell(w) > \ell(w_0) - \ell(wr) = \ell(w_0wr)$ thanks to Lemma 8 and the hypothesis. By Lemma 10, $B_r^- \subset B_{wr}^-$. On the other hand, using Lemma 11(b) we have:

$$\begin{aligned} B_{wr}^- &= B_{wr}^- \cap B = B_{wr}^- \cap (B_r^- B_r) \\ &= B_r^-(B_{wr}^- \cap B_r) \qquad \text{(use } B_r^- \subset B_{wr}^-\text{)} \\ &= B_r^-(B^r \cap B^{w_0 w} \cap B)^r\ . \end{aligned}$$

But Lemma 6(b) and the above calculation $\ell(w_0wr) < \ell(w_0w)$ together imply that $B^r \cap B^{w_0wr} \subset B$, so $B \cap B^{w_0w} \subset B^r$ and

$B^r \cap B^{w_0w} \cap B = B \cap B^{w_0w} = B_w^-$. □

<u>LEMMA 13</u>. $B = B_w^- B_w$ (<u>with</u> $B_w^- \cap B_w = H$).

<u>Proof</u>. Write $w = r_1 \dots r_k$ (reduced). Set $w_j = r_j \dots r_k$, so Lemma 8 shows that $\ell(w_0 w_j^{-1}) = \ell(w_0 w_{j+1}^{-1}) - 1$. This (and $w_0^2 = 1$) allows us to apply Lemma 12(b) repeatedly to obtain:

$$B = B_{w_0}^- = B_{r_k}^- (B_{w_0 w_k^{-1}}^-)^{w_k} = B_{r_k}^- (B_{r_{k-1}}^-)^{w_k} (B_{w_0 w_{k-1}^{-1}}^-)^{w_{k-1}}$$

$$= \dots = B_{r_k}^- \dots (B_{r_1}^-)^{w_2} (B_{w_0 w^{-1}}^-)^{w},$$ the last factor

being B_w. Then collapse the first k terms to B_w^- using 12(a): work inside parentheses from right to left. □

The preceding development follows Richen [1]. In order to conform with the convention of Borel [5], we set

$B_w' = B \cap B^{w_0 w^{-1}} \quad (= B_{w^{-1}}^-)$.

<u>THEOREM 3</u>. $G = \bigcup_{w \in W} B_w' w B$.

<u>Proof</u>. $G = \bigcup_{w \in W} BwB$ (by Theorem 1(a))

$= \bigcup_{w \in W} B_{w^{-1}}^- B_{w^{-1}} wB$ (by Lemma 13)

$= \bigcup_{w \in W} B_w' wB$, since $w^{-1} B_{w^{-1}} w \subset B$. □

<u>DEFINITION</u>. The BN-pair (G,B,N,R) is <u>split</u> if there exists a normal subgroup U of B with $U \cap H = e$, $B = H.U$.

<u>Example</u>. In GL(n,k) the upper triangular unipotent group plays the role of U.

<u>Assume henceforth that</u> (G,B,N,R) <u>is split</u>.

Evidently $U \cap U^{w_0} = e$. We set $U_w' = (B_w' \cap U) = (U \cap U^{w_0 w^{-1}})$. (In GL(n,k), U_r' is a "one-parameter" subgroup of U corresponding to the (i,i+1) position if $r = (i,i+1)$ in S_n.)

THEOREM 4 (Refined Bruhat Decomposition). If $\{s_w\} \subset N$ is a set of coset representatives for W, with $(s_w)^{-1} = s_{w^{-1}}$, then $G = \bigcup_{w \in W} U'_w s_w H\, U$, with uniqueness of expression.

Proof. Existence of the decomposition follows at once from Theorem 3. If $us_w b = u's_{w'}b'$ (u,u' in $U'_w, U'_{w'}$, $b,b' \in B$), then $w = w'$ (so $s_w = s_{w'}$) by Theorem 1(b). Then $s_{w'}^{-1}u'^{-1}us_w = b'b^{-1} \in B$, but the left side is in U^{w_0}, and $U^{w_0} \cap B = e$. □

Notation. If $g \in BwB$, we write $g = u_g s_w h_g v_g$ ($u_g \in U'_w, v_g \in U$, $h_g \in H$). Strictly speaking, we ought to write $s_{w(g)}$, but in practice we usually make use of this decomposition relative to a fixed $w \in W$.

12.4 Complements for GL_n

Specializing now to the case $G = GL(n,k)$ (k field), we obtain a few other properties which will be needed in §13. See Borel [5,§3].

(1) Consider the product map $U^- \times H \times U \xrightarrow{\phi} G$, where $U^- = U^{w_0}$ is the lower triangular unipotent group. Let $U^- \times U \xrightarrow{\psi} G$ be the restriction of ϕ to $U^- \times U$. ϕ (hence also ψ) is injective: $s_{w_0} U^- H\, U = U'_{w_0} s_{w_0} H\, U$, and uniqueness of expression in the latter (Theorem 4) implies uniqueness in the former as well. If $g \in G$, let $\Delta_i(g)$ = determinant of $i \times i$ submatrix in upper left corner of g. We claim that:

$$\begin{cases} \operatorname{Im}\phi = \{g \in G \mid \Delta_i(g) \neq 0,\quad i = 1,\dots,n\} \\ \operatorname{Im}\psi = \{g \in G \mid \Delta_i(g) = 1,\quad i = 1,\dots,n\} \\ \text{If } g \in \operatorname{Im}\phi,\ \Delta_i(g) = h_1 \dots h_i \quad (\text{where } h \text{ is the } H\text{-component of } g). \end{cases}$$

Proceed by induction on n. If $n^- \in U^-$, $b \in B$,

$$n^-.b = \left(\begin{array}{c|c} n' & 0 \\ \hline n'' & 1 \end{array}\right)\left(\begin{array}{c|c} b' & b'' \\ \hline 0 & b_{nn} \end{array}\right) = \left(\begin{array}{c|c} n'b' & n'b'' \\ \hline n''b' & b_{nn}+n''b'' \end{array}\right).$$

Then if $g \in G$ is written $\left(\begin{array}{c|c} g' & q \\ \hline p & g_{nn} \end{array}\right)$, $g = n^-b$ iff $b' = n'b'$,

$p = n''b'$, $q = n'b''$, $g_{nn} = b_{nn} + n''b''$. Under the assumption that $\Delta_i(g) \neq 0$ for all i, induction allows us to find n', b' with $g' = n'b'$. Since n', b' are nonsingular, n'', b'' are then (uniquely) determined, and finally $b_m = g_{nn} - n''b''$. So g can be written in the form $n^- b$, as required. The other assertions are now obvious (since det is multiplicative).

(2) In any split BN-pair one can define a notion of "root". Here we have a very concrete version, which has already figured tacitly in the discussion of Siegel sets. For $i \neq j$ we define a <u>root</u> $\alpha_{ij} : H \to k^*$ by $\alpha_{ij}(h) = h_i/h_j$. <u>Positive</u> (resp. <u>negative</u>) roots are those for which $i < j$ (resp. $i > j$). The $n-1$ roots $\alpha_i = \alpha_{i,i+1}$ $(i=1,\ldots,n-1)$ are called <u>simple</u>. Notice that the positive root α_{ij} just assigns to $h \in H$ the value by which conjugation by h multiplies the (i,j) entry in the one-parameter unipotent group $U'_{(i,i+1)}$.

Obviously each root can be written <u>uniquely</u> in the form $\prod_{i=1}^{n-1} \alpha_i^{m_i}$, with all $m_i \geq 0$ or all $m_i \leq 0$.

The Weyl group $W = S_n$ acts on roots in a natural way, viz., $w\alpha_{ij} = \alpha_{w(i),w(j)}$. This action satisfies: $(w\alpha)(h) = \alpha(h^w)$, w being viewed on the right side as a permutation matrix in G.

<u>Exercises</u>. (a) w_0 interchanges the positive and negative roots. (Is it true that $w_0\alpha = \alpha^{-1}$?)

(b) $\ell(w)$ = number of positive roots made negative by w. (This generalizes (a).)

Recall from 12.2 the description of parabolic groups containing B. The maximal (proper) parabolics BW_JB correspond to sets J of cardinality $n-2$. Let $P = BW_JB$ for $J = R - \{(\lambda,\lambda+1)\}$, $\lambda = 1,2,\ldots,n-1$. It is easy to describe the matrices $g \in P_\lambda$:

$$g = \left(\begin{array}{c|c} \lambda \times \lambda & \lambda \times (n-\lambda) \\ \hline 0 & (n-\lambda)\times(n-\lambda) \end{array}\right)$$

LEMMA. *If* $w \in W$ *lies in no proper parabolic subgroup* BW_JB, *then for any* j *there exists* i *such that* $w(\alpha_i) = \prod_{\ell=1}^{n-1} \alpha_\ell^{n_{i\ell}}$ *with* $n_{ij} < 0$ *(and, of course, all other* $n_{i\ell} \leq 0$).

Proof. The hypothesis means that $w \notin P_\lambda$ for all λ. In turn this means that for some index $a \leq \lambda$, $w(a) > \lambda$ (see the above description of P_λ). This forces existence of $b > \lambda$ with $w(b) \leq \lambda$, since w has a single nonzero entry in each row and each column. So $a < b$, and $w(\alpha_{a,b}) = \alpha_{w(a),w(b)} = \alpha_{w(b),w(a)}^{-1}$ $= (\alpha_{w(b)}\alpha_{w(b)+1} \cdots \alpha_{w(a)})^{-1}$. But $w(b) \leq \lambda < w(a)$, so α_λ^{-1} occurs on the right side. What we have shown is that for any λ, there is a positive root α_{ab} such that $w(\alpha_{ab})$ is negative and involves α_λ^{-1}. If α_{ab} is written as product of simple roots, obviously there must be one, say α_i, for which the same property holds. This is the lemma (for $j = \lambda$). □

§13. Siegel property (and applications)

We now return to the situation of §10, with $G = GL(n,\mathbb{R})$. The reader is urged to adapt the results below to $SL(n,\mathbb{R})$ as an exercise. We are following Borel [5, §4].

13.1 Siegel sets revisited

Recall that we defined a Siegel set $\Sigma_{t,u} = K A_t U_u$ in §10, as a certain closed set in G, and proved that for large enough t,u the translates of such a set by $\Gamma = GL(n,\mathbb{Z})$ will cover G. We find it convenient now to relax the definition, and to call Siegel set any set of the form $\Sigma = KA_t\omega$, where $\omega \subset U$ is relatively compact. (The exact bound on ω is not important in what follows.) In particular, if ω is chosen to be open, we get open Σ. It is trivial to deduce from §10 the existence of open Siegel sets Σ for which $G = \Sigma.\Gamma$.

Recall that a set is relatively compact iff its closure is compact. In U, which is homeomorphic to a euclidean space, this is equivalent to boundedness of the coordinates (in absolute value).

LEMMA 1. If $\omega \subset U$ is relatively compact, then $\bigcup_{a \in A_t} a\omega a^{-1}$ is relatively compact.

Proof. If $a \in A_t$, all $\alpha_i(a) \leq t$ (see 12.4 for definition of simple roots α_i), whence all $\alpha_{ij}(a) \leq t^N$ (e.g., for $N = n(n-1)/2$). In turn, $|(ana^{-1})_{ij}| = |\alpha_{ij}(a).n_{ij}| \leq C\, t^N$ if ω is bounded by C.□

LEMMA 2. $\omega \subset U^-$ is relatively compact iff $a(\omega)$ is relatively compact in A.

Proof. (=>) Clear.

($\Leftarrow$) If $x = k_x a_x n_x$, $x\, n_x^{-1} = k_x a_x$ runs over a relatively compact set $K.a(\omega)$ as x runs over ω. But (1) in 12.4 implies that $U^- \times U$ is homeomorphic (under the product map) to a closed subspace of G, so ω (= inverse image of relatively compact set) is relatively compact.□

From now on we always let the Weyl group representatives s_w be the obvious ones (the permutation matrices); for SL_n, a slight modification of this choice would be needed, however. In particular, $s_w \in K$.

LEMMA 3. If $g \in BwB$, then $a_g = a(s_w^{-1} u_g s_w)\, a(h_g)$. (This shows how to measure the discrepancy between the Iwasawa and Bruhat decompositions of the diagonal part.)

Proof.
$$\begin{aligned} g &= u_g s_w h_g v_g = s_w(s_w^{-1} u_g s_w) h_g v_g \\ &= s_w\, c\, h_g v_g \qquad (c = s_w^{-1} u_g s_w) \\ &= s_w k_c a_c n_c h_g v_g \,. \end{aligned}$$

But h_g has the obvious Iwasawa decomposition $k(h_g)\, a(h_g)$, where $k(h_g)$ is diagonal with entries ± 1 and hence commutes with A. So

$$\begin{aligned} g &= s_w k_c a_c h_g (h_g^{-1} n_c h_g) v_g \\ &= \underbrace{s_w k_c k(h_g)}_{K}\ \underbrace{a_c a(h_g)}_{A}\ \underbrace{(h_g^{-1} n_c h_g) v_g}_{U} \end{aligned}$$

Uniqueness in the Iwasawa decomposition now yields the lemma. □

Let $\Lambda\mathbb{R}^n$ be the exterior algebra of $\mathbb{R}^n$ (the tensor algebra modulo the ideal generated by all $v \otimes v$). The ith exterior power $\Lambda^i\mathbb{R}^n$ has canonical basis (decreed to be orthonormal) $e_{\ell_1} \wedge \ldots \wedge e_{\ell_i}$ $(\ell_1 < \ell_2 < \ldots < \ell_i \leq n)$, and euclidean norm $\|\ \|$. Of course $\Lambda^i\mathbb{R}^n = 0$ for $i > n$.

For $i \leq n$, $g \in G$, define $\Phi_i(g) = \|g(e_1 \wedge \ldots \wedge e_i)\|$. When $i = 1$, $\Phi_i = \Phi$ (in §10). The functions Φ_i provide a convenient way to establish bounds for subsets of G. If $\lambda_i(b) = b_{11} \ldots b_{ii}$ $(b \in B)$, it is clear that:

$$\Phi_i(g) = \Phi_i(a_g) = \lambda_i(a_g),$$

$$\Phi_i(gb) = \Phi_i(g) \quad |\lambda_i(b)| \quad (g \in G,\ b \in B).$$

LEMMA 4. $C \subset A$ <u>is relatively compact iff there exist</u> $\alpha, \beta > 0$ <u>such that</u> $\alpha \leq \Phi_i(c) \leq \beta$ $(c \in C,\ i = 1,\ldots,n)$. <u>If</u> $\det c = 1$ <u>for all</u> $c \in C$, <u>then</u> C <u>is relatively compact iff all</u> $\alpha_i(c)$ (α_i = <u>simple root) are bounded above and below.</u>

Proof. Exercise. □

LEMMA 5. <u>If</u> $g \in BwB$, $\Phi_i(g) = \Phi_i(a_g) = \Phi_i(s_w^{-1} u_g s_w)\ \Phi_i(h_g)$.

Proof. Lemma 3 (and the fact that Φ_i is multiplicative on A). □

LEMMA 6. (a) <u>If</u> $n \in U^-$, $\Phi_i(n) \geq 1$ <u>for all</u> i.

(b) <u>If</u> $g \in G$, $\Phi_i(a_g) \geq \Phi_i(h_g)$ <u>for all</u> i.

Proof. (a) Because of triangular form, $n(e_1 \wedge \ldots \wedge e_i)$ $= e_1 \wedge \ldots \wedge e_i$ + linear combination of other canonical (orthonormal) basis elements, so $\Phi_i(n) \geq 1$. Since $s_w^{-1} u_g s_w \in U^-$, Lemma 5 and (a) imply (b). □

LEMMA 7. <u>For each</u> i <u>there exists</u> $d_i > 0$ <u>such that</u> $\|g_v\|$ $\geq d_i \|v\| \Phi_i(g)$ <u>for all</u> g <u>in a given Siegel set</u> Σ <u>and all</u> $v \in \Lambda^i\mathbb{R}^n$.

Proof. It suffices to treat the case $\|v\| = 1$. Write $\Sigma = K A_t \omega$ (ω relatively compact in U). Then if $f_j = e_{\ell_1} \wedge \ldots \wedge e_{\ell_i}$ is a canonical basis element of $\wedge^i\mathbb{R}^n$, we have for $g \in \Sigma$,

$$a_g f_j = a_{\ell_1} \cdots a_{\ell_i} f_j \qquad (a = a_g)$$

$$= (a_1 \cdots a_i)\left(\frac{a_{\ell_1}}{a_1} \cdots \frac{a_{\ell_i}}{a_i}\right) f_j$$

$$= \lambda_i(a_g)\alpha_1(a_g)^{m_{1,j}} \cdots \alpha_{n-1}(a_g)^{m_{n-1,j}} f_j ,$$

where all $m_{kj} \leq 0$ because $\ell_1 < \ell_2 < \cdots < \ell_i$. But $g \in \Sigma$ forces $\alpha_k(a_g) \leq t$ for all k (i.e., $a_g \in A_t$). Therefore we get a <u>lower</u> bound $\delta > 0$ for the coefficient involving the roots. On the other hand, $g \in \Sigma$ implies that $n_g v = \Sigma\, \beta_j(n_g) f_j$ (f_j running over the canonical basis), with n_g in a relatively compact set; so there exists $\delta' > 0$ with $\|n_g v\|^2 = \Sigma\, \beta_j(n_g)^2 \geq \delta' > 0$ (since $\|v\| = 1$). Finally, $\|gv\|^2 = \|k_g a_g n_g.v\|^2 = \|a_g n_g.v\|^2 \geq \delta\lambda_i(a_g)^2(\Sigma\, \beta_j(n_g)^2)$ $\geq d_i^2\lambda_i(a_g)^2$ $(d_i^2 = \delta\delta') = d_i^2\Phi_i(g)^2$.□

13.2 <u>Fundamental sets and Siegel property</u>

We call subgroups T_1, T_2 of a group T <u>commensurable</u> if $T_1 \cap T_2$ has finite index in both T_1 and T_2. (<u>Exercise</u>. Verify that commensurability is an equivalence relation.)

With $\Gamma = GL(n,\mathbb{Z})$ as before, we call any subgroup Γ' of $GL(n,\mathbb{Q})$ commensurable with Γ an <u>arithmetic subgroup</u> of G. An important example is the <u>principal congruence subgroup</u> of level m, the kernel of the canonical homomorphism $GL(n,\mathbb{Z}) \to GL(n,\mathbb{Z}/m\mathbb{Z})$. (Why does this kernel have finite index in Γ ?) If Γ' is arithmetic, we say that a subset Ω of G is a <u>fundamental set</u> for Γ' in G (relative to K) iff

(F_0) $K\Omega = \Omega$,

(F_1) $\Omega\Gamma' = G$,

(F_2) (<u>Siegel property</u>) $b \in GL(n,\mathbb{Q})$ implies that $\{\gamma \in \Gamma' \mid \Omega b \cap \Omega\gamma \neq \emptyset\}$ is finite.

<u>Exercise.</u> (a) Show that (F_2) is equivalent to (F_2') : If C is a finite subset of $GL(n,\mathbb{Q})$, then $\{\gamma\in\Gamma' \mid \Omega C \cap \Omega C\gamma \neq \emptyset\}$ is

finite.

(b) Using (a), prove that if a fundamental set exists for Γ in G, then one exists for any other arithmetic subgroup.

(Remark. One can avoid reference to $\mathbb{Q}$ in this set-up by passing from $GL(n,\mathbb{Q})$ to the "commensurability group" of Γ in G: Borel [5, p.105].)

We proceed to prove that Siegel sets always satisfy (F_2); this will be deduced from the following theorem.

THEOREM (Harish-Chandra). *Let* Σ *be a Siegel set in* G. *Let* $M \subset G$ *be a subset satisfying:*

(i) $M = M^{-1}$,

(ii) *For each* i, *there exists* $C_i > 0$ *such that for all* $m \in M$, $\Phi_i(h_m) \geq C_i$. *Then* $M_\Sigma = \{m \in M \mid \Sigma \cap \Sigma m \neq \emptyset\}$ *is relatively compact in* G.

Note that h_m depends on the choice of coset representatives s_w for W, which we have fixed. The reader should think of M as being essentially Γ. We have to allow more general M only in order to treat the case $b \neq e$ in (F_2); in turn, this is needed only in order to obtain information about arithmetic subgroups other than Γ (see above exercise).

COROLLARY. *If* Σ *is any Siegel set in* G, *then the Siegel property* (F_2) *holds for* Γ *(hence for any other arithmetic subgroup* Γ').

Proof of corollary. We are given some $b \in GL(n,\mathbb{Q})$. Set $M = \Gamma b^{-1} \cup b\Gamma$. Clearly (i) is satisfied. Since Γ, $b\Gamma$, Γb^{-1} are closed, discrete subsets of G, Harish-Chandra's theorem will imply that M_Σ is finite (hence that Γ_Σ in (F_2) is finite) if we know that M also satisfies (ii). For this, notice that all denominators of coefficients of matrices in Γ or $b\Gamma$ or M are bounded above in absolute value. Since G is the union of *finitely* many cosets BwB, it suffices to verify (ii) for $M \cap BwB$ (w fixed). Writing

$m = u_m s_w h_m v_m$, $s_w^{-1} m = \underbrace{(s_w^{-1} u_m s_w)}_{c_m} h_m v_m \in U^- H\, U$. It is clear that the denominators for the matrices $c_m h_m v_m$ are bounded above in absolute value, since this was true for m. Using (1) of 12.4, $\det^{i \times i}(c_m h_m v_m) = (h_m)_{11} \ldots (h_m)_{ii}$ (upper left $i \times i$ minor), so these rational numbers (of absolute value $\Phi_i(h_m)$) have denominators bounded above in absolute value.□

COROLLARY. *For $t \geq 2/\sqrt{3}$, $u \geq 1/2$, $\Sigma_{t,u}$ is a fundamental set for Γ in G.*

Proof. Use the preceding corollary and Theorem 10.3. □

13.3 *Proof of Harish-Chandra's theorem*

Since s_w is a permutation matrix, it is clear that s_w normalizes A. Note too that multiplication by a positive scalar leaves Σ invariant. Now we can proceed with the proof.

(1) *As m ranges over M_Σ , a_m, u_m, h_m range over relatively compact sets.*

Write $xm = y$ $(x, y \in \Sigma,\ m \in M_\Sigma)$. Apply Lemma 7 successively to v (using the fact that $xm \in \Sigma$) and to $m(e_1 \wedge \ldots \wedge e_i)$ (using the fact that $x \in \Sigma$), where $v \in \wedge^i \mathbb{R}^n$ is arbitrary:

$$\|(xm).v\| \geq d_i \|v\| \Phi_i(xm) ,$$

$$\|x.m.(e_1 \wedge \ldots \wedge e_i)\| \geq d_i \Phi_i(m)\ \Phi_i(x) .$$

Since the left side of the second inequality is $\Phi_i(xm)$, these combine to yield:

$$\|(xm)v\| \geq d_i^2 \|v\|\ \Phi_i(m)\ \Phi_i(x) .$$

In particular, if $v = (m^{-1})(e_1 \wedge \ldots \wedge e_i)$, then

$$\Phi_i(x) \geq d_i^2 \Phi_i(m^{-1})\ \Phi_i(m)\ \Phi_i(x),$$

hence $\Phi_i(m)\Phi_i(m^{-1}) \leq \dfrac{1}{d_i^2}$. (*)

On the other hand, condition (ii) of the theorem, along with Lemma 6, yields:

$$\Phi_i(m) = \Phi_i(a_m) \geq \Phi_i(h_m) \geq C_i > 0 . \quad (**)$$

Since $M = M^{-1}$, (*) and (**) together imply that $\Phi_i(a_m)$, $\Phi_i(h_m)$ are bounded above (as well as below) on M_Σ. In view of Lemma 4, $a(M_\Sigma)$ and $h(M_\Sigma)$ are relatively compact.

Now Lemma 3 shows that $a(s_w^{-1}u_m s_w)$ traces a relatively compact set for $m \in M_\Sigma \cap BwB$. But $s_w^{-1}u_m s_w \in U^{-}$, so Lemma 2 implies that the set of all $s_w^{-1}u_m s_w$ (or the set of all u_m) here is also relatively compact. G being a finite union of double cosets BwB, this proves that $u(M_\Sigma)$ is relatively compact.

(2) With these preliminaries disposed of, we proceed by induction on n (the case $n = 1$ being trivial). The reader may find it helpful to follow through in detail the case $n = 2$, where the argument below is simpler.

The basic strategy of the proof is to show that each set $M_\Sigma \cap BwB$ is relatively compact. So we may fix $w \in W$ and exploit the Bruhat decomposition. Two cases arise, the first requiring a direct argument and the second using induction.

Case (a): s_w lies in no maximal parabolic group P_λ.

Take $m \in M_\Sigma$, so $xm = y$ for some $x, y \in \Sigma$. Since (as we remarked above) Σ is invariant under positive scalars, we can adjust this equation so that $|\det x| = 1$. Let $X = \{x \in \Sigma \mid |\det x| = 1$ & $xm \in \Sigma$ for some $m \in M \cap BwB\}$, and $Y = \{x \in \Sigma \mid y = xm$ for some $x \in X$, $m \in M \cap BwB\}$. It evidently suffices to show that X and Y (or just $a(X)$, $a(Y)$) are relatively compact.

If $y = mx$ ($x \in X$, $y \in Y$, $m \in M \cap BwB$), then we can write $k_y a_y n_y = k_x a_x n_x u_m s_w h_m v_m = k_x s_w k_c a_c n_c (s_w^{-1} a_x s_w) h_m v_m$ $[c = s_w^{-1} a_x n_x u_m a_x^{-1} s_w]$. But $s_w^{-1} a_x s_w \in A$ (as remarked at the outset), so we can rearrange the right side of the equation (as in the proof of Lemma 3) to get $a_y = a_c . (s_w^{-1} a_x s_w) . a(h_m)$. (***)

We showed in (1) that u_m traces a relatively compact set; since

$x \in \Sigma$, n_x traces a relatively compact set also. Therefore (Lemma 1), $a_x(n_x u_m)a_x^{-1}$ traces a relatively compact set. Then c, a_c do likewise, and because of (***) it <u>will suffice to prove that</u> $a(X)$ <u>is relatively compact</u>.

If $x \in \Sigma$, all $\alpha_j(a_x)$ are bounded above. But $1 = |\det x| = \det a_x = \Phi_n(a_x)$, so it will suffice to show that all $\alpha_j(a_x)$ are bounded <u>below</u> (see Lemma 4). We know that a_c and $a(h_m)$ have coefficients bounded above and below (by the preceding paragraph and (1)), while $y \in \Sigma$ implies all $\alpha_i(a_y)$ are bounded above. So for each i, (***) shows that $(w\alpha_i)(a_x) = \alpha_i(a_x^{s_w}) = \alpha_i(s_w^{-1}a_x s_w)$ is bounded <u>above</u>. By our hypothesis on w and the lemma (2) of 12.4, we see that for any j there exists i with $w\alpha_i = \prod \alpha_\ell^{m_{i\ell}}, m_{ij} < 0$, all $m_{i\ell} \leq 0$. All $\alpha_\ell(a_x)$ being bounded above, we conclude from this that $\alpha_j(a_x)$ cannot be arbitrarily small.

<u>Case (b):</u> s_w <u>lies in some</u> P_λ .

Notice that $s_w \in P_\lambda$ forces $BwB \subset P_\lambda$, so it <u>suffices to show that each</u> $M_\Sigma \cap P_\lambda$ is <u>relatively compact</u> $(\lambda = 1,\ldots,n-1)$.

Now if $m \in M_\Sigma \cap P_\lambda$, $xm = y$ $(x,y \in \Sigma)$, we can multiply both sides by k_x^{-1} and use the fact that $AU \subset P_\lambda$ to conclude that:

$$M_\Sigma \cap P_\lambda = \{m \in M \mid \Sigma \cap P_\lambda \text{ meets } (\Sigma \cap P_\lambda)m\} .$$

To apply induction, we need a more precise description of P_λ (cf. 12.4 (2)), for which the reader should fill in details: $P_\lambda = S.R$ (semidirect product in topological sense), where S is the direct product $GL(\lambda,\mathbb{R}) \times GL(n-\lambda,\mathbb{R})$ and S normalizes the unipotent group $R = \left(\begin{array}{c|c} I_\lambda & * \\ \hline 0 & I_{n-\lambda} \end{array}\right)$. Here $S = S_1 \times S_2$ consists of the matrices $\left(\begin{array}{c|c} GL_\lambda & 0 \\ \hline 0 & GL_{n-\lambda} \end{array}\right)$. Notice that $K \cap P_\lambda = K \cap S = \left(\begin{array}{c|c} O_\lambda & 0 \\ \hline 0 & O_{n-\lambda} \end{array}\right) = K_1 \times K_2$.

Let π, π_0, π_1, π_2 be the respective projections of P_λ onto

S, R, S_1, S_2. The reader can quickly verify that π_1, π_2 transform the Iwasawa parts in P_λ to those in S_1, S_2, hence transform Σ into Siegel sets Σ_1. Σ_2. Set $M_i = \pi_i(M \cap P_\lambda)$, so $\pi_i(M_\Sigma \cap P_\lambda) \subset (M_i)_{\Sigma_i}$. It is equally obvious that π_1, π_2 transform the Bruhat decomposition in P_λ to that in S_1, S_2. The reader can check that the hypotheses (i), (ii) of the theorem hold for M_i, Σ_i in S_i. By induction, we see that $(M_i)_{\Sigma_i}$, and hence $\pi_i(M_\Sigma \cap P_\lambda)$, are relatively compact for $i = 1,2$.

It remains to show that $\pi_0(M_\Sigma \cap P_\lambda)$ is relatively compact, i.e., bounded. If $xm = y$ ($m \in M_\Sigma \cap P_\lambda$, $x,y \in \Sigma$), we saw above that we can assume $x,y \in P_\lambda$. So the following makes sense:

$$k_x a_x n_x m = k_y a_y n_y ,$$

$$\underbrace{k_x a_x \pi(n_x)\pi(m)}_{S} \quad \underbrace{\pi(m)^{-1} \pi_0(n_x)\pi(m) \ \pi_0(m)}_{R} = \underbrace{k_y a_y \pi(n_y)}_{S} \quad \underbrace{\pi_0(n_y)}_{R} .$$

Therefore, $\pi_0(n_y) = \pi(m)^{-1}\pi_0(n_x)\pi(m)\pi_0(m)$. $\pi_0(n_x)$, $\pi_0(n_y)$ are bounded, since x,y are in a Siegel set (and since $\pi(n_x)$, $\pi(n_y)$ are bounded), whereas $\pi(m)$ was shown above to be bounded. So the same is true of $\pi_0(m)$. $\square$

13.4 Finite presentation of Γ

From the existence of a fundamental domain for Γ having good connectedness properties we can deduce that Γ is finitely presentable. In the case at hand this can also be shown in purely group-theoretic ways and one can even give explicit presentations with very few generators and relations. The approach of Borel and Harish-Chandra works equally well, however, for arithmetic subgroups of other linear algebraic groups, where detailed group-theoretic information is not so readily available.

It will be convenient to work in the space $X = K\backslash G$; we know that it is possible to choose an open "Siegel set" in X which is "fundamental" for Γ in an obvious sense (cf. 13.2), and we call such a set Ω. Finally, denote by Δ the finite set $\Gamma_\Omega = \{\gamma \in \Gamma : \Omega\gamma \cap \Omega \neq \emptyset\}$. Obviously Ω is connected, as is X.

THEOREM. (a) Γ is generated by the finite set Δ.
(b) The finite subgroups of Γ form finitely many conjugacy classes in Γ (in fact, each such group has a conjugate contained in Δ).

Proof. (a) Let Γ' be the subgroup of Γ generated by Δ, and let X' be the open set $\Omega\Gamma'$. If $\Omega\gamma \cap \Omega\gamma' \neq \emptyset$ for $\gamma' \in \Gamma'$, then clearly $\gamma'\gamma^{-1} \in \Delta \subset \Gamma'$ and $\gamma \in \Gamma'$. So if $\Gamma'' = \Gamma - \Gamma'$, the open set $X'' = \Omega\Gamma''$ is disjoint from X'. But $X = X' \cup X''$ and X is connected, so $X'' = \emptyset$, $\Gamma = \Gamma'$.
(b) If Γ' is a finite subgroup of Γ, then Γ' fixes some point $x \in X$. (The reader should verify this as an exercise, using for example the characterization of X in 11.2.) Now $x = x_0\gamma$ for some $x_0 \in \Omega$, $\gamma \in \Gamma$. Clearly this forces $x_0 \in \Omega \cap \Omega\gamma\Gamma'\gamma^{-1}$, $\gamma\Gamma'\gamma^{-1} \subset \Delta$. The assertions of (b) follow at once. □

We turn now to the question of relations for Γ. We adapt some of Behr's reasoning (see Behr [1]), retaining the above notation. Notice that X is homeomorphic to the (semidirect) product AU. Since each of A and U has the topology of a euclidean space (group structures being ignored, cf. 11.1), X is simply connected.

Let F be the free group on a set $\overline{\Delta}$ in 1-1 correspondence with Δ. Let L be the normal subgroup of F generated by the local relations:

$$\overline{\delta}_1\overline{\delta}_2\,\overline{\delta_1\delta_2}^{-1} = e ,$$

whenever $\delta_1, \delta_2, \delta_1\delta_2 \in \Delta$. Since Δ is finite, the number of local relations is finite, so the group $H = F/L$ is finitely presented. Let

$$F \xrightarrow{s} H \xrightarrow{t} \Gamma$$

be the canonical epimorphisms, so $ts(\overline{\delta}) = \delta$ for $\overline{\delta} \in \overline{\Delta}$.

THEOREM. t is an isomorphism; so Γ is finitely related (in fact, the local relations suffice).

Proof. The idea will be to construct a covering for X and use the simple connectedness of X. We want a commutative diagram

$$\begin{array}{ccc} \Omega \times H & \xrightarrow{(1,t)} & \Omega \times \Gamma \\ q \downarrow & & \downarrow \text{product map} \\ S & \xrightarrow[p]{} & X \end{array}$$

Let H have the discrete topology and Ω its subspace topology. If $\Omega \times H$ is given the product topology, then $(1,t)$ is continuous. To obtain S we define a relation on $\Omega \times H$ by $(x,h) \sim (x',h')$ iff there exists $\delta \in \Delta$ with $x' = x\delta$, $h = s(\overline{\delta})h'$. This relation is symmetric because Δ is inverse closed and $s(\overline{\delta})^{-1} = s(\overline{\sigma^{-1}})$ in H (local relations). The relation is transitive (again by the local relations): $(x',h') \sim (x'',h'')$ yields $\delta_0 \in \Delta$ with $x'' = x'\delta_0$, $h' = s(\overline{\delta_0})h''$, which combine to give $x'' = x(\delta\delta_0)$ (hence $\delta\delta_0 \in \Delta$), $h = s(\overline{\delta})s(\overline{\delta_0})h'' = s(\overline{\delta}\overline{\delta_0})h'' = s(\overline{\delta\delta_0})h''$. By definition, S is the quotient of $\Omega \times H$ by this equivalence relation; we give S the quotient topology (so q is open) and write $q(x,h) = \langle x,h \rangle$.

Next let $p(\langle x,h \rangle) = x.t(h)$. This makes the diagram commute. (Check that p is well defined.) Clearly p must be continuous and surjective.

We observe next that if $\Psi = q(\Omega \times \{e\})$, then $\Omega \times \{e\} \xrightarrow{q} \Psi \xrightarrow{p} \Omega$ is the identity, so p maps Ψ homeomorphically onto Ω. H acts on S (on the right) in the obvious way and $\Psi H = S$. We leave it as an exercise for the reader to prove that S is connected, and hence that (S,p) is a covering of X (since each point of X lies in a translate of the open set Ω, which is evenly covered by a corresponding union of translates of Ψ). [To prove S connected, show that any translate of Ψ is connected to Ψ by a finite chain of pairwise overlapping H-translates of Ψ.]

Because X is simply connected, p must therefore be 1-1. Now $t(h) = e$ implies that $p\langle x,h\rangle = x$ (for all $x \in \Omega$), or $(x,h)\sim(x,e)$ (all $x \in \Omega$) since p is 1-1. So there exists $\delta \in \Delta$ with $h = s(\overline{\delta})e$, forcing $e = t(h) = \delta$, or $h = s(e) = e$. □

Remark. The function field case involves some further complications. See for example Stuhler [1].

13.5 Corners and arithmetic groups

Borel, Serre [1] have used the main results of reduction theory to study the cohomology of arithmetic groups. Here we indicate very briefly what they do, taking for G a semisimple group such as $SL(n,\mathbb{R})$ and for Γ an arithmetic subgroup such as $SL(n,\mathbb{Z})$. Denote by X the quotient of G by a maximal compact subgroup (e.g., the upper half-plane when $G = SL(2,\mathbb{R})$). The idea is to embed X in a "manifold with corners" $\overline{X}$. Set-theoretically, $\overline{X}$ is the union of X and of euclidean spaces corresponding to the various proper parabolic subgroups of G (countably many lines in the case of $SL(2,\mathbb{R})$) Topologically, X is the interior of the Hausdorff space $\overline{X}$, and the boundary has the homotopy type of the Tits "building" of G (constructed from the collection of parabolic subgroups, cf. 15.4 below for a p-adic analogue). By combining the two main results of reduction theory (cf. 10.3 and 13.2 above), Borel and Serre deduce [1, Theorem 9.3] that Γ operates properly on $\overline{X}$ and that the quotient $\overline{X}/\Gamma$ is compact. The cohomology of the manifold then yields precise information about the cohomology of Γ.

V. GL_n AND SL_n (p-ADIC AND ADELIC GROUPS)

In §14 we take a brief look at GL_n and SL_n from the adelic point of view. Here some arithmetic themes recur in a noncommutative setting: class number, strong approximation, fundamental domains. The p-adic groups which contribute to the adele group are very interesting in their own right; in §15 we indicate in the case of SL_2 what sorts of discrete and compact subgroups one can expect to find.

§14. Adelic groups

The basic reference for this section is Borel [2]. K denotes a number field with integers $\mathcal{O}_K$ and completions K_v. (More generally, K could be any global field in most of what follows.)

14.1 Adelization of a linear group

If a closed subgroup of $GL(n,\mathbb{C})$ is defined by K-polynomial conditions (such as $\det(g)-1 = 0$) we can also consider the subgroup of $GL(n,K_v)$ defined by the same conditions. Important examples are (cf. Introduction): GL_n, SL_n, O_n, Sp_n, additive group $\left\{\begin{pmatrix} 1 & t \\ 0 & 1 \end{pmatrix}\right\}$, multiplicative group GL_1. One can also work with the orthogonal group of an arbitrary quadratic form on K^n; in this case, the study of the corresponding quadratic forms over completions K_v has been an important technique for a long time. To study all completions simultaneously we resort again to the use of adeles.

Let $G = GL_n$, SL_n, etc., so that G_K, G_v $(= G_{K_v})$ have an obvious meaning. For finite v, we can also define $G_{\mathcal{O}_v}$, the subgroup of G_v consisting of matrices with coefficients in $\mathcal{O}_v$ and determinant a unit in $\mathcal{O}_v$. Since G_v is topologized as a subset of some n^2-space over K_v, we see that G_v is locally compact while $G_{\mathcal{O}_v}$ (when defined) is an open compact subgroup. Therefore we may form the restricted topological product of the G_v relative to the $G_{\mathcal{O}_v}$ (see 3.1). With componentwise group operation, the resulting object is a locally compact group (denoted G_A), the adele group of G (cf. Weil [1]). We already made this construction in two cases: G = additive group, $G_A = \mathbb{A}_K$; G = multiplicative group, $G_A = \mathbb{J}_K$. As in those cases, we can view the topology on G_A as determined by the requirement that for each finite $S \supset S_\infty$, $G_{A(S)} = \prod_{v \in S} G_v \times \prod_{v \notin S} G_{\mathcal{O}_v}$

(with product topology), should be an open subgroup. When $S = S_\infty$, we write $G_{A(\infty)}$; this group will play an important role below.

Evidently we may embed G_K in G_A diagonally as a discrete subgroup. (Examples: $K \hookrightarrow A_K$, $K^* \hookrightarrow J_K$.)

14.2 Class number

Let $c(G)$ = number of double cosets $G_{A(\infty)} \backslash G_A / G_K$. Borel [2] proves that $c(G)$ is finite for all cases of interest in this theory. When G is the additive group, we already know that $A_K = A_K(\infty) + K$ (see 4.2), which means $c(G) = 1$. When G is the orthogonal group of a non-degenerate quadratic form on $\mathbb{Q}^n$, $c(G)$ turns out to be the number of "classes" in the "genus" of the form in question; that this number is finite is a classical theorem (cf. O'Meara [1, §103]). (For further study of class numbers, see Platonov, Bondarenko, Rapinčuk [1].)

THEOREM. Let $G = GL_n$. Then $c(G)$ = class number of K.

(Exercise. In case $n = 1$, deduce this from our earlier work on ideles.)

Before giving the proof of the theorem we have to recall some facts about lattices; references will be given to § 81 of O'Meara's book. An (arithmetic) lattice in K^n (resp. $K_v{}^n$) is a finitely-generated O_K (resp. O_v) submodule containing a vector space basis of K^n (resp. $K_v{}^n$) and contained in the O_K (resp. O_v) span of some basis of K^n (resp. $K_v{}^n$). For example, when $n = 1$, L is a lattice in K iff L is a fractional ideal. In the case of a PID such as O_v, the structure theorem for finitely-generated modules implies that an arithmetic lattice is the same thing as a lattice in the usual sense: the O_v-span of a basis of $K_v{}^n$.

Lattices L, L' are isomorphic iff there exists $g \in GL(n,K)$ (resp. $g \in GL(n,K_v)$) such that $L = L'g$. (We write the group action on the right to conform with Borel's notation.) For example, when

$n = 1$, two lattices (fractional ideals) in K are isomorphic iff they belong to the same ideal class. In particular, not all lattices in K^n need be isomorphic. On the other hand, since O_v is a PID, all lattices in ${K_v}^n$ are isomorphic.

Given a lattice L in K^n, its O_v-span in K_v^n is obviously a lattice, which we call the <u>localization of</u> L <u>at</u> v and denote by L_v.

<u>LEMMA 1.</u> <u>If</u> L <u>is a lattice in</u> K^n, $L = \bigcap_{v \text{ finite}} (K^n \cap L_v)$.

<u>Proof.</u> See O'Meara, p. 218.

<u>LEMMA 2.</u> <u>If</u> L <u>is a lattice in</u> K^n, $L_v = O_v^n$ <u>for almost all</u> v.

<u>Proof.</u> Give a direct proof (exercise) or see O'Meara, p. 218.

<u>LEMMA 3.</u> <u>Given lattices</u> $L_v \subset K_v^n$ (v <u>finite) with almost all</u> $L_v = O_v^n$, <u>then</u>

$$\bigcap_{v \text{ finite}} (K^n \cap L_v) \quad \text{\underline{is a lattice in} } K^n.$$

<u>Proof.</u> O'Meara (81:14).

<u>LEMMA 4.</u> <u>Isomorphism classes of lattices in</u> K^n <u>correspond 1-1 with ideal classes of</u> K.

<u>Proof.</u> O'Meara (81:5).

<u>Proof of theorem.</u>

(1) Define an action of G_A on the set of lattices in K^n by:

$$L \cdot g = \bigcap_{v \text{ finite}} (K^n \cap L_v g_v) \quad .$$

By Lemma 2 (and the definition of G_A) almost all $L_v g_v = O_v^n$, so Lemma 3 implies that $L \cdot g$ is a lattice in K^n. Clearly $L \cdot (g_1 g_2) = (L \cdot g_1) \cdot g_2$.

(2) If $g \in G_K \hookrightarrow G_A$, the action just defined is the usual one: $(Lg)_v = L_v g = L_v g_v$, so $L \cdot g = \bigcap_v (K^n \cap (Lg)_v) = Lg$ (by Lemma 1).

(3) The action of G_A is <u>transitive</u>: Set $L_0 = O_K^n$ (standard lat-

tice), let L be any lattice. Using the fact that all lattices in K_v^n are isomorphic, find $g_v \in GL(n,K_v)$ such that $L_v = (L_0)_v g_v$ $(= O_v^n g_v)$ for each v. Lemma 2 implies that $(L_0)_v = L_v$ $(= O_v^n)$ for almost all v, so we set $g_v = e \in G_{O_v}$ for all these v. Similarly, set $g_v = e$ for v infinite. Then $g = (g_v)$ is a well defined element of G_A, and in view of Lemma 1 we have $L = L_0 \cdot g$.

(4) <u>The stabilizer of L_0 is $G_{A(\infty)}$</u>: Clearly, $G_{A(\infty)}$ does stabilize L_0. On the other hand, if the stabilizer of L_0 is a larger group, then its projection on some G_v must properly contain G_{O_v}. But L_0 (hence O_v^n) lies in the intersection of all images of O_v^n under this subgroup of G_v, whereas G_{O_v} is precisely the stabilizer of O_v^n -- a contradiction.

(5) Combining the previous steps, we see that $G_{A(\infty)} \backslash G_A$ may be identified with the set of all lattices in K^n, in such a way that right translation by G_K in $G_{A(\infty)} \backslash G_A$ corresponds to the usual action of G_K on lattices. Therefore $c(G)$ = number of isomorphism classes of lattices = class number of K, by Lemma 4. □

14.3 <u>Strong approximation</u>

We return to the situation of 14.1 (so G is arbitrary). If $S \supset S_\infty$ is a finite set of primes, we write $G^{(S)} = \prod_{v \in S} G_v$ and view this as a subgroup of G_A by letting all components outside S be e. If $S = S_\infty$, we write $G^{(\infty)}$. [This notation is not standard in the literature; we adopt it to avoid confusion with $G_{A(S)}$, $G_{A(\infty)}$.]

G is said to have the <u>strong approximation property</u> relative to S (or absolute strong approximation, for $S = S_\infty$) if

$$\overline{G^{(S)}\ G_K} = G_A$$

(equivalently, if G_K is dense in the restricted product of the G_v, $v \notin S$). The requirement that $S \supset S_\infty$ can be omitted (cf. §6) if one makes certain adjustments in proofs, but we shall not do so here. Therefore: absolute strong approximation implies strong approximation relative to any S.

LEMMA. If G has (absolute) strong approximation, then $c(G) = 1$.

Proof. If U is an open subgroup of G_A including $G^{(\infty)}$ (e.g., $U = G_{A(\infty)}$), then for arbitrary $g \in G_A$, the open set Ug must meet the dense set $G^{(\infty)}G_K$, whence $g \in UG^{(\infty)}G_K = UG_K$, or $G_A = UG_K$. In particular, $G_A = G_{A(\infty)}G_K$, or $c(G) = 1$. □

COROLLARY (modulo Theorem below). $c(SL_n) = 1$.

Remark. Kneser [2] (and Behr [4]) have developed necessary conditions for G to have the strong approximation property: G must be "simply connected," $G^{(S)}$ noncompact, the "radical" of G unipotent (so GL_n does not qualify). These conditions also turn out to be sufficient, as Kneser [3] and Platonov [2] have shown (cf. Margulis [4] and Prasad [3] for the function field case).

For the following theorem we follow the proof in Moore [1, Lemma 13.1]. First we recall that for any field F, $SL(n,F)$ is generated by its elementary unipotent subgroups $X_\alpha = \{x_\alpha(t), t \in F\}$, where $x_\alpha(t)$ has 1's on the diagonal and the single nonzero entry t off the diagonal (in the position corresponding to the "root" α: cf. 12.4). This is really just a fact from linear algebra. Now each X_α is a copy of the additive group of the field. The idea of the proof below is to apply strong approximation (for the additive group) to each X_α and thereby get strong approximation for SL_n.

THEOREM. $G = SL_n$ has (absolute) strong approximation.

Proof. ($S = S_\infty$) The embedding of X_α in G is evidently "functorial", i.e., we get $X_{\alpha,A} \hookrightarrow G_A$ and $X_{\alpha,v} \hookrightarrow G_v$ in a natural way. By strong approximation for the additive group, $X_{\alpha,v}$ (α root, $v \notin S$) lies in the closure of G_K in the restricted product of $\{G_w, w \notin S\}$ (even in the closure of $X_{\alpha,K}$). But, as remarked above, the $X_{\alpha,v}$ (α root) generate G_v, so G_v lies in the closure of G_K in the indicated restricted product. (Here we view each G_v as a subgroup of G_A in the obvious way.) Therefore all G_v lie in the closure of $G^{(\infty)}G_K$, as do all their finite products $\prod G_v$. But the latter are evidently dense in G_A, so $G_A = \overline{G^{(\infty)}G_K}$.□

14.4 Reduction theory

Since G_K is a discrete subgroup of the locally compact group G_A, it is natural to consider fundamental domains, compactness or noncompactness of G_A/G_K, finiteness of Haar measure, harmonic analysis, etc.

A subset Ω of G_A may be called a fundamental set for G_K if

(i) $$G_A = \Omega G_K \quad ,$$

(ii) $$\Omega\Omega^{-1} \cap G_K \text{ is finite.}$$

It turns out that most of the work in defining a suitable Ω has already been done in Chapter IV above, for $G = GL_n$ or SL_n .

THEOREM. If Σ is a fundamental set for $GL(n,\mathbb{Z})$ in $GL(n,\mathbb{R})$ then $\Omega = \Sigma \times \prod_p GL(n,\mathbb{Z}_p)$ is a fundamental set for $GL(n,\mathbb{Q})$ in $GL(n,\mathbb{A}_{\mathbb{Q}}) = G_A$.

Proof. This is left as an exercise. (The reader may wish to compare our discussion of fundamental domains in the adele and idele groups.) Here we have $c(GL(n,\mathbb{Q}))$ = class number of $\mathbb{Q} = 1$, so $G_A = G_{A(\infty)}G_{\mathbb{Q}}$. This fact makes the proof of the theorem easy.

Remarks.

(1) The case of a number field $K \supsetneq \mathbb{Q}$ can be reduced to the case $K = \mathbb{Q}$ by "restriction of scalars"; so for reduction theory only the latter is essential. (See Borel [2, 1.4].)

(2) One can define a subgroup G_A^0 (analogous to J_K^0 in the case GL_1) and give a criterion for G_A^0/G_K to be compact, generalizing the result for ideles. The reader might want to attempt this generalization (cf. Gel'fand, Graev, Pyatetskii-Shapiro [1, p.379]).

§15. SL_2 (over p-adic fields)

In this section we look closely at $SL(2,K)$ in the nonarchimedean local case. This is only the first step in an extensive development begun by Iwahori-Matsumoto [1] and completed by Bruhat-Tits [1], [2], cf. Tits [4], Hijikata [1]. But some of the main techniques are already present in what follows. The object is to get information about compact and discrete subgroups of $SL(2,K)$, analogous to what one has for $SL(2,\mathbb{R})$ or $SL(2,\mathbb{C})$ (the archimedean local cases). This is especially interesting for harmonic analysis (cf. Gel'fand, Graev, Pyatetskii-Shapiro [1], Macdonald [1], and the work of Harish-Chandra), but has group-theoretic consequences as well.

One could also ask about fundamental domains for discrete subgroups in $SL(2,K)$, as we did in the real case. But here the situation changes considerably. Whereas we found noncompact fundamental sets of finite Haar measure in $SL(2,\mathbb{R})$ (relative to $SL(2,\mathbb{Z})$), there is a theorem in the p-adic case which asserts: If Γ is a discrete subgroup of $G = SL(2,\mathbb{Q}_p)$ (for example) with G/Γ of finite invariant measure, then G/Γ is compact. (See Serre [5, II, 1.5], Tamagawa [1].) The nature of discrete subgroups is not altogether clear in the p-adic case, but Ihara's theorem (proved below) gives some further insight. (See also Borel, Harder [1] and Prasad [4].) Notice that $SL(2,\mathbb{Z}_p)$ is not discrete (instead, it's compact)!

For our purposes it is enough to assume that K is a field with discrete valuation v, ring of integers $\mathcal{O}$, unique maximal ideal $P = \pi\mathcal{O}$ (π a "prime" element), $k = \mathcal{O}/P$. (In the Bruhat-Tits theory one soon has to assume that K is complete, but we shall not do so.) Example: $\mathbb{Q}_p$, $\mathbb{Z}_p$, $p\mathbb{Z}_p$, $k = \mathbb{F}_p$. Since we are working with 2×2 matrices, some of the routine calculations will be left for the reader.

15.1 Infinite dihedral group

It is well known that a finite group generated by two non-commuting involutions a, b is dihedral of order $2m$, if m = order of ab.

Concretely, one thinks of a (say) as a reflection and of b as this same reflection followed by a rotation of $2\pi/m$ (so the dihedral group is the group of rigid motions of a regular m-sided polygon).

On the other hand, if ab is allowed to have infinite order, we get the <u>infinite dihedral group</u> $D_\infty = \langle a,b;\ a^2 = b^2 = e\rangle$. It is an easy exercise to show that any infinite group generated by two non-commuting involutions is isomorphic to D_∞.

In particular, let N = group of monomial matrices in $G = SL(2,K)$ (as in §12), and let H be the subgroup of diagonal matrices $\begin{pmatrix} a & 0 \\ 0 & a^{-1} \end{pmatrix}$, with $a \in \mathcal{O} - \mathcal{P}$ (group of units of K).

(<u>Exercise</u>. H is normal in N.) Let $W = N/H$ (which is a group, by the exercise). If $w_0 = \begin{pmatrix} 0 & -1 \\ 1 & 0 \end{pmatrix}$ and $w_1 = \begin{pmatrix} 0 & \pi^{-1} \\ -\pi & 0 \end{pmatrix}$, then the respective images a, b in W obviously have order 2. Moreover, $w_0w_1 = \begin{pmatrix} \pi & 0 \\ 0 & \pi^{-1} \end{pmatrix}$ has infinite order modulo H, so by the previous remarks $\langle a,b\rangle \cong D_\infty$. On the other hand, $W = \langle a,b\rangle$: Monomial matrices are of the types $\begin{pmatrix} a & 0 \\ 0 & a^{-1} \end{pmatrix}$ and $\begin{pmatrix} 0 & a^{-1} \\ -a & 0 \end{pmatrix}$, $a \in K^*$. Modulo units, the first type is gotten from some power of w_0w_1; but we know from §12 (or directly) that the diagonal group has index 2 in N, so N is certainly generated by w_0 and w_0w_1, modulo H.

<u>CONCLUSION</u>: $W \cong D_\infty$. (This W will play the role of Weyl group in a BN-pair for G.)

15.2 <u>Lattices in</u> K^2

As observed in 14.2, the fact that $\mathcal{O}$ is a PID guarantees that all lattices in K^2 are isomorphic (under the action of $GL(2,K)$, which we apply on the <u>right</u>). Let e_1, e_2 be the standard basis of K^2, $L_0 = [e_1,e_2]$ the corresponding standard lattice $\mathcal{O}e_1 + \mathcal{O}e_2$.

Write $L \sim L'$ iff $L = L'g$ for some <u>scalar</u> matrix g. Since each element of K^* is product of a unit and a power of π, $L \sim L'$ iff $L = \pi^i L$, for some $i \in \mathbb{Z}$. In general, if L has stabilizer

S in G, then Lg has stabilizer $S^g = g^{-1}Sg$. So in particular, $L \sim L'$ implies that L, L' have the same stabilizer.

Since $G = SL(2,K)$ commutes with scalars, G acts on equivalence classes of lattices. It is easy to see that G has exactly two orbits, represented by (the classes of) $L_0 = [e_1, e_2]$, $L_1 = [\pi e_1, e_2]$. (Check!) It is not difficult to compute the respective stabilizers P_0, P_1 in G: $P_0 = SL(2,\mathcal{O})$ is obvious. If $\begin{pmatrix} a & b \\ c & d \end{pmatrix} \in G$ stabilizes L_1, then in particular $\pi ae_1 + \pi be_2 = \pi a'e_1 + b'e_2$ for some $a', b' \in \mathcal{O}$, hence $a \in \mathcal{O}$, $b \in \pi^{-1}\mathcal{O}$. Similarly we deduce $c \in \pi\mathcal{O}$, $d \in \mathcal{O}$; and then a, b, c, d can be given arbitrarily (with $ad-bc = 1$) satisfying these conditions. Therefore $P_1 = \left\{\begin{pmatrix} a & b \\ c & d \end{pmatrix} \in G \mid a,d \in \mathcal{O},\right.$ $\left. b \in \pi^{-1}\mathcal{O},\ c \in \pi\mathcal{O}\right\}$. What is $P_0 \cap P_1$? This group (call it B) is obviously $\left\{\begin{pmatrix} a & b \\ c & d \end{pmatrix} \in G \mid a,d \in \mathcal{O} - P,\ b \in \mathcal{O},\ c \in \pi\mathcal{O}\right\}$.

15.3 <u>BN-pair in G</u>

With N as in 15.1 and B as in 15.2, we want to verify the BN-axioms (§12). All but (BN1) are straightforward.

(1) <u>G is generated by B and N.</u>

We know (§12) that G is generated by N along with the upper triangular unipotent group. Matrices $\begin{pmatrix} 1 & b \\ 0 & 1 \end{pmatrix}$, $b \in \mathcal{O}$, already lie in B. To get arbitrary $b \in K$, we must multiply by negative powers of π; but conjugating $\begin{pmatrix} 1 & \pi b \\ 0 & 1 \end{pmatrix} \in B$ by $\begin{pmatrix} \pi^{-1} & 0 \\ 0 & \pi \end{pmatrix} \in N$ yields $\begin{pmatrix} 1 & \pi^{-1}b \\ 0 & 1 \end{pmatrix}$, so it is clear how to proceed.

(2) <u>$B \cap N$ is normal in N.</u>

Indeed, it is evident that $B \cap N$ is the group H described in 15.1, which is quickly seen to be normal in N.

(3) <u>$W = N/H$ is generated by involutions a, b.</u>

We saw in 15.1 that $W \cong D_\infty$ is generated by the cosets of $w_0 = \begin{pmatrix} 0 & -1 \\ 1 & 0 \end{pmatrix}$ and $w_1 = \begin{pmatrix} 0 & \pi^{-1} \\ -\pi & 0 \end{pmatrix}$ which have order 2 (mod H). <u>From now on we abuse notation by viewing w_0 and w_1 as generators of W.</u>

(4) <u>(BN2) holds.</u>

This is easy to verify; for example, $\begin{pmatrix}1 & 1\\ 0 & 1\end{pmatrix}^{w_0} \notin B$.

(5) It remains only to verify axiom (BN1); this will follow from Lemma 4, 5 below. Let U, V be (respectively) the upper and lower triangular unipotent subgroups of G. Then $U_{\mathcal{O}}$, U_{P}, etc. have an obvious meaning.

LEMMA 1. $B = V_P H\, U_{\mathcal{O}}$ ($= U_{\mathcal{O}} H\, V_P$), with uniqueness of expression.

Proof. $\begin{pmatrix}a & b\\ c & d\end{pmatrix} \overset{?}{=} \begin{pmatrix}1 & 0\\ g & 1\end{pmatrix}\begin{pmatrix}h & 0\\ 0 & h^{-1}\end{pmatrix}\begin{pmatrix}1 & f\\ 0 & 1\end{pmatrix} = \begin{pmatrix}h & hf\\ hg & hfg+h^{-1}\end{pmatrix}$ iff $a = h$, $b = hf$, $c = hg$, $d = hfg+h^{-1}$. These equations have (unique) solution $h = a$, $f = ba^{-1}$, $g = ca^{-1}$, and then $h \in \mathcal{O} - P$, $f \in \mathcal{O}$, $g \in P$; $d = hfg+h^{-1} = bca^{-1}+a^{-1}$ is automatic, since $ad-bc = 1$. □

LEMMA 2. For $i = 0,1$, $P_i = B \cup Bw_iB$.

Proof. First look at P_0. Since $P_0 = SL(2,\mathcal{O})$, the canonical map $\mathcal{O} \to \mathcal{O}/P \cong k$ induces a homomorphism $\phi : P_0 \to SL(2,k)$. (Although we don't need to know it, ϕ is actually surjective; why?) Clearly $\operatorname{Ker}\phi \subset B$ (indeed, B is precisely the inverse image of the upper triangular group, the "Borel group" in the usual BN-pair for $SL(2,k)$). Notice that ϕ sends w_0 to $\begin{pmatrix}0 & -1\\ 1 & 0\end{pmatrix} \in SL(2,k)$, which represents the nontrivial generator for the usual Weyl group in $SL(2,k)$. Using the Bruhat decomposition in the (rank 1) group $SL(2,k)$ and the fact that $\operatorname{Ker}\phi \subset B$, we immediately get $P_0 = B \cup Bw_0B$ by lifting back to P_0.

Now the matrix $g = \begin{pmatrix}0 & 1\\ \pi & 0\end{pmatrix} \in GL(2,K) - SL(2,K)$ normalizes B (check!), sends L_0 to L_1, hence conjugates P_0 to P_1, and $g^{-1}w_0g = w_1$. So $P_0 = B \cup Bw_0B$ forces $P_1 = B \cup Bw_1B$. □

Remark. The reader might try to verify this last point directly to see the advantage of the method of proof we have used.

LEMMA 3. Conjugation by w_0 sends $U_{\mathcal{O}}$ into $V_{\mathcal{O}}$, V_P into $U_{\mathcal{O}}$. Conjugation by w_1 sends $U_{\mathcal{O}}$ into V_P, V_P into $\pi^{-1}U_{\mathcal{O}}$.

Proof. This is an easy exercise.

LEMMA 4. *If* $\ell(w_i w) > \ell(w)$ *in* W, $Bw_iBwB = Bw_iwB$.

Proof. It suffices to prove $\subseteq$. There are two parallel cases to consider.

(i = 0) The assumption evidently means that a reduced expression for w (in terms of the generators w_0, w_1) begins on the left with w_1. Using Lemmas 1, 3 we therefore have:

$$Bw_0BwB = Bw_0(V_P H\, U_O)wB = BV_P^{w_0} H^{w_0}\, w_0\, U_O wB$$

$$= Bw_0 w(w^{-1}U_O w)\, B \subset Bw_0 wB$$

(since conjugation by w sends U_O into V_P into U_O into V_P ... by Lemma 3, and at each step the result is in B).

(i = 1) Here w begins with w_0 on the left. Using the symmetric version of Lemma 1, with Lemma 3,

$$Bw_1BwB = Bw_1(U_O\, H\, V_P)\, wB = BU_O^{w_1} H^{w_1}\, w_1\, V_P wB$$

$$= Bw_1 w(w^{-1}V_P\, w)B \subset Bw_1wB\ ,$$

since w sends $V_P \to U_O \to V_P \to \ldots \subset B$. □

LEMMA 5. *If* $\ell(w_i w) < \ell(w)$, $Bw_iBwB = Bw_iwB \cup BwB$.

Proof. (i = 0) Write $w = w_0w'$, $\ell(w') < \ell(w)$ by hypothesis. Then $\ell(w_0w') > \ell(w')$, allowing us to conclude from Lemma 4 that $Bw_0Bw'B = BwB$. Begin as in the proof of Lemma 4:

$$Bw_0BwB = Bw_0(V_P\ H\ U_O)w_0w'B$$

$$= BV_O\ w'B \quad \subset \quad Bw'B \cup Bw_0Bw'B \qquad \text{(by Lemma 2)}$$

$$= Bw_0wB \cup BwB \qquad \text{(by above use of Lemma 4)}.$$

(Here we have exploited the fact that $BV_O \subset P_0$, so Lemma 2 applies.) To complete the proof, notice that $Bw_0BwB = BV_O\ w'B$ must meet (hence contain) both double cosets on the right, because BV_O properly contains B.

(i=1) Argue similarly, using Lemmas 2 and 4. □

THEOREM. (G,B,N,R), $R = \{w_0H, w_1H\}$, *is a BN-pair.*

Exercise. Prove that the BN-pair just constructed is "saturated" (12.3).

Remark. It should be clear in principle how to carry out a similar program for SL(n,K); the reader may want to attempt this (cf. Iwahori-Matsumoto [1]). N should again be the full group of monomial matrices, B the inverse image in $SL(n,\mathcal{O})$ of the standard Borel group in SL(n,k). The Weyl group $W = N/B \cap N$ will be generated by n involutions, of which n-1 are the generators in $SL(n,\mathcal{O})$ of the usual Weyl group S_n. W is semidirect product of S_n and a free abelian group of rank n-1 (on which S_n acts), i.e., W is the so-called affine Weyl group associated with the ordinary Weyl group S_n.

COROLLARY. B, P_0, P_1 are the sole proper subgroups of G containing B, and each is self-normalizing (equal to its normalizer).

Proof. Use the characterization of parabolic groups in BN-pairs (12.2). □

As usual, a subset X of G is called bounded if $v(x_{ij}) \leq$ constant for all $x \in X$. This means simply that the coefficients of matrices in X do not involve arbitrarily high negative powers of π. In particular, G is not bounded, but P_0, P_1 clearly are bounded.

COROLLARY. P_0, P_1 are (non-conjugate) maximal bounded subgroups of G.

When K is assumed locally compact (§2), X is bounded iff X is relatively compact. Therefore, P_0, P_1 are maximal compact subgroups in that case.

Exercises. (a) Prove that a subset X of G is bounded iff X lies in a finite union of double cosets BwB.

(b) Let K be locally compact. Prove that G has an "Iwasawa decomposition" $G = P_0.A.U$, with $A = \begin{pmatrix} \pi^i & 0 \\ 0 & \pi^{-i} \end{pmatrix}$, $i \in \mathbb{Z}$. Can something similar be done using the maximal compact subgroup P_1? (Cf. Macdonald [1, II].)

15.4 Building attached to BN-pair

In his study of BN-pairs attached to finite simple groups, Tits was led to introduce certain geometric complexes on which the groups act, so-called "buildings". (See Bourbaki [2, Chapter 4, exercises].) In the case at hand we get a graph, which will now be described explicitly. In essence, the vertices are <u>lattices</u> in K^2 (and we could proceed from this point of view, suppressing the BN-structure, but we prefer to use the general set-up).

Define a graph I (cf. Appendix below) to have as set of vertices $G/P_0 \cup G/P_1$ and as edges $\{gP_0, gP_1\}$ $(g \in G)$. Notice that G acts transitively on the set of all edges, and the stabilizer of $\{P_0, P_1\}$ is $P_0 \cap P_1 = B$; therefore the edges correspond 1-1 with elements of G/B. The edges are also called <u>chambers</u> of the <u>building</u> I, and $C = \{P_0, P_1\}$ is the canonical chamber. A subgraph A (the canonical <u>apartment</u>) is defined by

$$
\cdots \quad \underset{w_0w_1P_0}{\circ} \text{---} \underset{w_0P_1}{\circ} \text{---} \underset{\underset{\shortparallel}{P_0}}{\circ} \text{---} \underset{\underset{\shortparallel}{P_1}}{\circ} \text{---} \underset{\underset{\shortparallel}{w_1P_0}}{\circ} \text{---} \underset{w_1w_0P_1}{\circ} \text{---} \circ \quad \cdots
$$
$$
\qquad\qquad\quad \underset{w_0P_0}{} \quad \underset{w_1P_1}{} \quad \underset{w_1w_0P_0}{}
$$

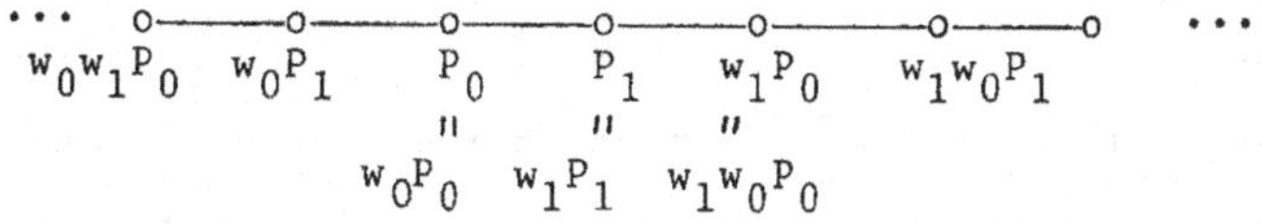

and the translates gA are called apartments of I. Clearly an apartment is a <u>tree</u> (cf. Appendix); it will be shown below that I is itself a tree. Notice that G acts on I <u>without inversion</u>, so $g \in G$ fixes an edge iff it fixes both endpoints. This means that G preserves some orientation of I; the most natural orientation is

$$\underset{gP_0}{\circ} \text{---} \!\!\rightarrow \underset{gP_1}{\circ} \quad .$$

Recall from 15.2 that P_0, P_1 represent the two conjugacy classes of stabilizers of (equivalence classes of) lattices in K^2. Therefore the vertices of I may be thought of as (classes of) lattices, with gP_i corresponding to $L_i g^{-1}$.

THEOREM. I is a tree.

Proof. We must prove that I is connected and has no circuits.
(1) To prove I connected, it suffices to show that any pair of chambers lies in some apartment. Since G acts transitively on chambers, we can assume we are given $\{P_0,P_1\}$ and $\{gP_0,gP_1\}$. But $G = BNB$ with $B \subset P_i$, so we can write $g = bw$. Then both chambers obviously lie in bA.
(2) Again using transitivity of G on chambers, we can suppose I has a circuit including C:

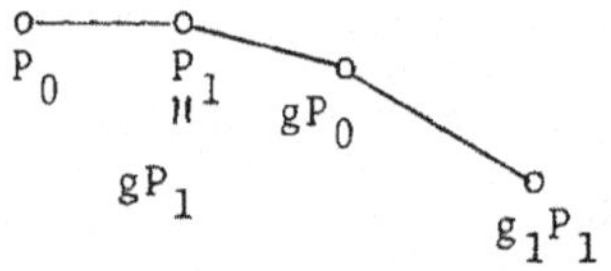

Now $gP_0 \neq P_0$ (the circuit involves no reversals) but $gP_1 = P_1$. Since $P_i = B \cup Bw_iB$, this forces $g \in Bw_1B$, so we can assume $g = bw_1$ (because $B \subset P_i$). Proceed to the next vertex g_1P_1 (so $g_1P_0 = gP_0 = bw_1P_0$). Again $g_1P_1 \neq gP_1$, forcing $g_1^{-1}bw_1 \in P_0$, $g_1^{-1}bw_1 \notin P_1$, hence $g_1^{-1}bw_1 \in Bw_0B$. Therefore we may write $g_1 \in Bw_1Bw_0B = Bw_1w_0B$ (see Lemma 4 in 15.3), or equally well $g_1 = b'w_1w_0$ (since $B \subset P_i$). Continuing in this way, we see that if the circuit is completed, we will have some $b_0w_1w_0w_1\cdots$ in $P_0 = B \cup Bw_0B$, contradicting the uniqueness of the Bruhat decomposition. □

Exercise. With C,A as above, show that B, N may be characterized by

$$B = \{g \in G \mid gC = C\}$$
$$N = \{g \in G \mid gA = A\}.$$

15.5 Ihara's theorem; maximal compact subgroups

Now we can obtain some information about certain discrete or compact subgroups of G.
(1) First, we obtain a theorem of Ihara [1] (using an approach sug-

gested by Serre [5, II, 1.5].

THEOREM. Let Γ be a subgroup of G containing no nontrivial bounded subgroup. Then Γ is free.

Proof. We know that G, hence Γ, acts on I without inversion. If some $\gamma \neq e$ in Γ fixes a vertex, say gP_0, then $g^{-1}\gamma g \in P_0$, so the (nontrivial) subgroup of Γ generated by γ is bounded, contrary to hypothesis. We conclude that Γ acts freely on I; since I is a tree, a well known theorem (see Appendix) states that Γ is free. □

COROLLARY (Ihara). Let K be locally compact. If Γ is a discrete, torsion-free subgroup of G, then Γ is free.

Proof. If Γ had a nontrivial bounded (i.e., relatively compact) subgroup, this subgroup would be both discrete and bounded, hence finite, contradicting the assumption that Γ is torsion-free. Therefore the theorem applies. □

Remarks. Serre [3] and Borel-Serre [1] have made effective use of the building of a BN-pair to investigate the discrete subgroups of both real and p-adic Lie groups, such as SL_2. Their methods lead to an upper bound on the cohomological dimension of certain torsion-free discrete groups, which is 1 in the case of $SL(2,\mathbb{R})$ or $SL(2,\mathbb{Q}_p)$. From the Stallings-Swan theorem on groups of cohomological dimension ≤ 1 one can then deduce that the group in question is free. Although the real and p-adic cases lead to rather different sorts of buildings, the algebraic approach furnished by BN-pairs does allow one to unify such results as the theorem of Ihara and the classical group-theoretic result that a torsion-free subgroup of $SL(2,\mathbb{Z})$ is free. It might also be pointed out that in the case of arithmetic groups (e.g. $SL(2,\mathbb{Z})$), Selberg proved that there always exists a torsion-free subgroup of finite index (see Borel [5, §17] for a very general theorem of this type).

(2) Next we consider bounded subgroups of G (one of the chief motives for the Bruhat-Tits theory).

In a tree (such as I) any two vertices can be joined by a path, and if reversals are forbidden there is obviously only one such path (otherwise we could form a circuit). The number of edges in such a path is called the distance between the vertices. Obviously a group acting on a tree must preserve distances.

We claim that each orbit in I under a bounded subgroup of G is itself bounded. For this we have to look more closely at how G acts on I. There are two types of vertices, and G acts transitively on those of each given type. So if we want to compute distances it suffices to begin with the vertex P_0 or P_1. According to the proof of Theorem 15.4, any two vertices lie in a common apartment; so it suffices to be able to compute distances in A in order to bound distances in I. Clearly, the distance between P_0 and wP_0 (resp. P_0 and wP_1) in A is approximately $\ell(w)$. (What is the precise distance?) Now if C is a bounded subgroup of G, it lies in a finite union of double cosets BwB, thanks to the exercise of 15.3. So the orbit of a vertex (such as P_0) under C is clearly bounded.

THEOREM. *Each bounded subgroup of* G *lies in a conjugate of* P_0 *or* P_1 *(so there are precisely two conjugacy classes of maximal bounded subgroups in* G).

Proof. It suffices to prove that if C is a bounded subgroup of G, then C fixes some point of I. Let X be an arbitrary C-orbit, and let X' be the union of X and those vertices which lie along minimal paths joining points of X. By our previous remarks, X (hence X') is bounded, and of course X' is connected. If d is the largest distance between vertices in the set X', we proceed step-by-step to reduce d. Evidently C stabilizes X'; the "extreme" points of X' (points at distance d from some point of X') are permuted by C, so if we discard them we will have a C-stable subset of X' having "diameter" 2 less than d. At the stage $d \leq 1$ we have an edge or a vertex stable under C, but in the former case the ends must also be fixed. Therefore C does fix some vertex of I. □

COROLLARY. *If* K *is locally compact, every compact subgroup of* G *lies in a conjugate of* P_0 *or* P_1.

(For a more direct treatment in the case of GL_n, using the action of G on lattices, see Serre [1, LG 4.30].)

Appendix. Graphs and free groups.

We shall sketch briefly some facts about graphs; a good treatment of these matters can be found in Serre [5, I, §2-3].

(1) A (combinatorial) *graph* X consists of a set of *vertices* and a set of *edges* (pairs of vertices), with an edge of the form  permitted. If one specifies origin and terminus of each edge, one has a notion of oriented (or directed) graph. X is *connected* if each pair of vertices can be joined by a path. A *circuit* (of length n) is a subgraph of the form

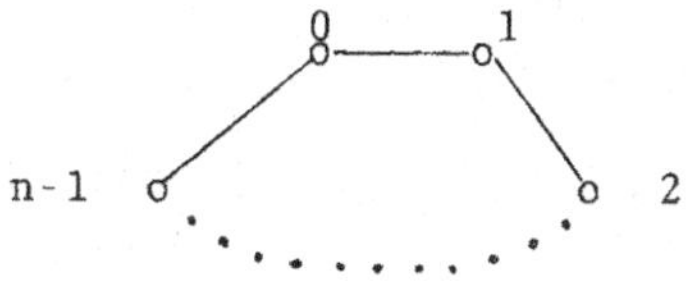

(with n distinct vertices $0, 1, 2, \ldots, n-1$). X is called a *tree* if X is connected and has no circuits.

(2) Given a group G and a subset S, one can form a graph $\Gamma(G,S)$ with vertices G and (directed) edges $\underset{g}{\circ}\!\longrightarrow\!\underset{gs}{\circ}$ $(g \in G,\ s \in S)$. G acts on this graph, preserving the orientation. It is easy to verify that:

$\Gamma(G,S)$ is connected iff S generates G;

$\Gamma(G,S)$ has circuit of length 1 iff $e \in S$;

$\Gamma(G,S)$ has no circuit of length 1 or 2 iff $S \cap S^{-1} = \emptyset$.

For example, if G is cyclic of order n on $S = \{s\}$, we have $\Gamma(G,S)$ = (directed) circuit of length n. If $G = D_\infty$, $S = \{a,b\}$ (see 15.1), $\Gamma(G,S)$ looks like:

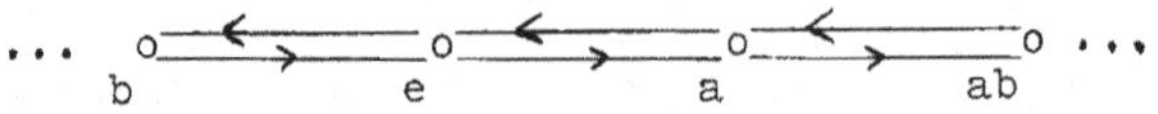

(3) A group G *acts* on a graph X if there is a homomorphism $G \to \operatorname{Aut} X$. In particular, G sends edges to edges, paths to paths, etc. G *acts freely* on X if G acts without inversion (i.e., no element of G reverses endpoints of an edge) and if no $g \neq e$ fixes any vertex of X. (Example: G acts freely on $\Gamma(G,S)$.)

THEOREM. (a) *If* G *is the free group on a set* S, *then* $\Gamma(G,S)$ *is a tree (on which* G *acts freely)*

(b) *If* $S \subset G$ *and* $\Gamma(G,S)$ *is a tree, then* G *is free on* S.

The proofs are almost obvious: connectedness of the graph signifies that S generates G, while absence of a circuit signifies absence of relations among the generators S.

Exercise. Let G be free on S. Let T be a tree on which G acts trivially. Show how to form a tree, whose vertices are pairs of vertices from T, $\Gamma(G,S)$, in such a way that G acts freely on this tree.

(4) THEOREM. *Let* X *be a tree,* G *a group acting freely on* X. *Then* G *is free.*

(As an immediate corollary, one gets *Schreier's theorem*: *A subgroup of a free group is free.*)

We indicate briefly how the proof goes (cf. the above exercise):

(i) Form the *quotient graph* X/G (vertices are the G-orbits of vertices in X; two G-orbits are endpoints of an edge if two vertices in the respective orbits are endpoints of an edge in X), let $\pi: X \to X/G$ be the projection. Of course, X/G need not be a tree.

(ii) Use Zorn's Lemma to find a maximal tree T' in X/G.

(iii) Using just the fact that X/G is a connected graph, verify that vert T' = vert X/G. (Otherwise, could construct a larger tree.)

(iv) Use Zorn's Lemma to find a tree T in X maximal with respect to the property that π maps T injectively into T'.

(v) Verify that $\pi(T) = T'$, so T is a tree of representatives of the G-orbits in X.

(vi) Since G acts freely on X, the elements $g \in G$ correspond 1-1 with the translates gT (which are pairwise disjoint, and which together cover X).

(vii) Form a graph $\overline{X}$ with vertices gT $(g \in G)$; let gT, $g'T$ be endpoints of an edge if some element of gT is joined to some element of $g'T$ by an edge. Since G acts freely on X, X can be assumed to have an orientation preserved by G, so we may endow $\overline{X}$ with corresponding orientation (preserved by G, acting in the obvious way on $\overline{X}$).

(viii) $\overline{X}$ is a tree: $\overline{X}$ is obviously connected. If a circuit exists, construct a (possibly longer) circuit in X by adding path segments in various of the trees gT, thereby contradicting the fact that X is a tree.

(ix) If $S = \{g \neq e \mid$ some edge of X begins in T and ends in $gT\}$, verify that $\overline{X} \cong \Gamma(G,S)$.

(x) By Theorem (b) in (3), G is free on S.

VI. THE CONGRUENCE SUBGROUP PROBLEM

In this chapter we look more deeply into the arithmetic structure of such linear groups as $SL(n,\mathbb{Q})$. It is well known that for almost all fields K, $SL(n,K)$ is close to being a simple group: the only proper normal subgroups lie in the finite center consisting of scalar matrices. (The only exceptions involve very small n and K.) We might be tempted to ask how close $SL(n,O_K)$ comes to being simple, when O_K is the ring of algebraic integers of a number field K.

Take $\Gamma = SL(n,\mathbb{Z})$, for example. There are a lot of obvious normal subgroups here, arising as kernels of "reduction mod q" maps (q any positive integer): $SL(n,\mathbb{Z}) \to SL(n,\mathbb{Z}/q\mathbb{Z})$. We denote the kernel by $\Gamma_q = \{x \in \Gamma \mid x \equiv 1 \pmod{q\mathbb{Z}}\}$ and call it the principal congruence subgroup of level q. Clearly $[\Gamma : \Gamma_q] < \infty$, since $SL(n,\mathbb{Z}/q\mathbb{Z})$ is a finite group. Call a subgroup of Γ a congruence subgroup if it includes some Γ_q. The Congruence Subgroup Problem is this: Is every subgroup of finite index in Γ a congruence subgroup? We could broaden the question to $G = SL(n,\mathbb{Q})$ by redefining "congruence subgroup" to mean "arithmetic subgroup including some Γ_q". Then we are asking: Is every arithmetic subgroup of G a congruence subgroup? (Even if the answer is yes, it would still have to be asked whether there are any non-obvious normal subgroups in Γ of infinite index.) The answer is in fact no when $n = 2$, but yes when $n \geq 3$. Although this may seem odd at first, the smaller group $SL(2,\mathbb{Z})$ has a vast assortment of normal subgroups of finite index, only a "few" of which are congruence subgroups. (This has been known since the work of F. Klein and others in the 19th century.)

When $\mathbb{Q}$ is replaced by an arbitrary number field K and $\mathbb{Z}$ by the ring of integers of K, the problem still makes sense. For $SL(n,K)$, $n \geq 3$, the answer remains yes unless K is totally imaginary, i.e., has no real embeddings. If $SL(n,K)$ is replaced by another simple simply connected algebraic group of "split" type (symplectic group, spin group, etc.), the congruence subgroup problem again has a natural formulation, and the answer is the same: yes, unless K is totally imaginary. Placed in its most natural setting - say that of simple algebraic groups having relative rank at least 2 over global fields - the problem has not yet been completely solved.

Our goal here is quite modest: to introduce some of the basic techniques (due to Matsumoto, Moore, and others) which are applicable to split groups of rank at least 2, while limiting ourselves to the special case $SL(n,\mathbb{Q})$. We shall first reformulate the problem as a question about central extensions, then show how this leads inexorably to the deep arithmetic results of Moore. While we treat the elementary group-theoretic portion of the problem in some detail, we will only be able to sketch the theory of Moore, which requires a lengthy development of its own. In doing this much, we hope to make the literature of the subject more accessible to the reader.

§16. Reformulation of the problem

16.1 Topological groups

We have already used some of the familiar properties of topological groups in earlier chapters. We shall also need some less familiar properties, cf. Higgins [1, Chapter II].

By "neighborhood" of a point x in a topological space we always mean an arbitrary set containing an open set to which x belongs. Since left (or right) translation in a topological group is a homeomorphism, it is enough to specify the neighborhoods of the identity element e. To be precise, we say that a collection F of open neighborhoods of e in a topological group G is a fundamental system of open neighborhoods of e if each neighborhood of e contains some member of F. The following properties of F are then readily established:

(F1) $U, V \in F$ $\Rightarrow$ there exists $W \in F$ such that $W \subset U \cap V$,

(F2) $a \in U \in F$ $\Rightarrow$ there exists $V \in F$ such that $Va \subset U$,

(F3) $U \in F$ $\Rightarrow$ there exists $V \in F$ such that $V^{-1}V \subset U$,

(F4) $U \in F$, $a \in G$ $\Rightarrow$ there exists $V \in F$ such that $a^{-1}Va \subset U$.

Conversely, given an abstract group G and a nonempty collection F of subsets of G, each containing e, which satisfies (F1)-(F4), there exists a unique topology on G such that G is a topological group having F as a fundamental system of open neighborhoods of e. The idea of the proof is to specify a basis $\mathcal{B}$ for the topology of G: $\mathcal{B} = \{Ua \mid U \in F,\ a \in G\}$. It has to be shown that the union of $\mathcal{B}$ is G and that for $B, B' \in \mathcal{B}$, $B \cap B'$ is a union of sets from $\mathcal{B}$; then $\mathcal{B}$ is a basis for a unique topology on G whose open sets are arbitrary unions from $\mathcal{B}$.

An important special case arises when F consists of subgroups of G. Then the requirements (F1)-(F4) boil down to the following:

(G1) $A, B \in F$ $\Rightarrow$ there exists $C \in F$ such that $C \subset A \cap B$,

(G2) $A \in F, g \in G \Rightarrow$ there exists $B \in F$ such that $g^{-1}Bg \subset A$.

These requirements are met, for example, by any collection of subgroups of G containing all conjugates of its members and closed under pairwise intersections. Then the open sets of G are all unions of right (or left) cosets of the given subgroups.

Examples. (1) Take F to be a chain of normal subgroups of G. For a prime p, the subgroups $p^n\mathbb{Z}$ of $\mathbb{Z}$ define in this way a topological group structure on $\mathbb{Z}$; the topology is of course just the familiar p-adic topology.

(2) Take F to be the collection of all subgroups of finite index in G. It is easy to see that $A \cap B$ and $g^{-1}Ag$ again have finite index, if $A, B \in F$.

16.2 Subgroup topologies on $SL(n,\mathbb{Q})$ and $SL(n,\mathbb{Z})$

Let $G = SL(n,\mathbb{Q})$, $\Gamma = SL(n,\mathbb{Z})$. Recall that a subgroup of G is called arithmetic if it is commensurable with Γ. In particular, the arithmetic subgroups included in Γ are just the subgroups of finite index in Γ, which according to 16.1 form a fundamental system of open neighborhoods of 1 for a topological group structure on Γ.

If q is a positive integer, we have defined the principal congruence subgroup of level q to be the kernel Γ_q of the reduction mod q map. An arithmetic subgroup of G is called a congruence subgroup if it includes some Γ_q.

Our aim now is to verify conditions (G1)-(G2) of 16.1 for the collection of arithmetic (resp. congruence) subgroups of G.

First consider intersections. If H, H' are arithmetic subgroups of G, all the following indices are finite: $[\Gamma:H \cap \Gamma]$, $[\Gamma:H' \cap \Gamma]$, $[H:H \cap \Gamma]$, $[H':H' \cap \Gamma]$. But intersections of subgroups of finite index are again of finite index. It follows that $[\Gamma:H \cap H' \cap \Gamma]$ and $[H \cap H':H \cap H' \cap \Gamma]$ are finite. So $H \cap H'$ is arithmetic. If moreover $H \supset \Gamma_p$, $H' \supset \Gamma_q$, then $H \cap H' \supset \Gamma_p \cap \Gamma_q$. But $\Gamma_p \cap \Gamma_q$ clearly equals Γ_r, where r is the least common multiple of p and q. So the intersection of congruence subgroups is another such.

Next consider conjugates. We claim that for $g \in G$, the associated inner automorphism $\sigma: x \mapsto gxg^{-1}$ takes arithmetic (resp. congruence) subgroups of G to groups of the same type. We may view σ as an automorphism of the full algebra of $n \times n$ matrices over $\mathbb{Q}$, with $\sigma(x)_{ij} = Q_{ij}(x_{11},\ldots,x_{nn})$ a polynomial over $\mathbb{Q}$ having zero constant term. Take $q \in \mathbb{Z}$ to be a common denominator for all coefficients of the various Q_{ij} (and the corresponding polynomials associated with σ^{-1}). Note that $(\sigma(x)-1)_{ij} = (\sigma(x-1))_{ij} = Q_{ij}(x_{11}-1,\ldots,x_{nn}-1)$; similarly for σ^{-1}. By our choice of q, we have $\sigma(\Gamma_q) \subset \Gamma$, $\sigma^{-1}(\Gamma_q) \subset \Gamma$. Combined with the preceding note, this shows that Γ_r includes $\sigma^{-1}(\Gamma_{rq})$ for any r.

Now let H be an arithmetic subgroup of G, so H is commensurable with Γ_q. In turn, $\sigma(H)$ is commensurable with $\sigma(\Gamma_q)$. But $\sigma(\Gamma_q)$ is arithmetic: Clearly $\sigma^{-1}(\Gamma_q) \cap \Gamma_q$ has finite index in $\sigma^{-1}(\Gamma_q)$. So $\Gamma_q \cap \sigma(\Gamma_q)$ has finite index in Γ_q, forcing $\sigma(\Gamma_q) = \Gamma \cap \sigma(\Gamma_q)$ to have finite index in Γ.

Suppose moreover that $H \supset \Gamma_r$. By our choice of q, $\Gamma_r \supset \sigma^{-1}(\Gamma_{rq})$. Applying σ, we obtain: $\sigma(H) \supset \sigma(\Gamma_r) \supset \Gamma_{rq}$, whence $\sigma(H)$ is also a congruence subgroup.

To sum up, we now have two subgroup topologies on G: the arithmetic topology T_a and the congruence topology T_c (which is included in T_a). The Congruence Subgroup Problem becomes: Is $T_a = T_c$?

The congruence topology on G is actually a familiar one (in disguise). As a moment's thought should make clear, it is just the topology induced on G by its diagonal embedding in $G^{(\infty)} = SL(n, \mathbb{A}^f)$, cf. 14.3. Here we denote by $\mathbb{A}^f$ the ring of "finite adeles", the restricted product of the fields $\mathbb{Q}_p$ relative to the subrings $\mathbb{Z}_p$ (with the infinite prime omitted).

16.3 Review of topology

We have to recall some further facts from topology which have special bearing on the behavior of topological groups. First we assemble a few properties of subgroups.

PROPOSITION. Let H be a subgroup of a topological group G. Then:

(a) $\bar{H}$ is a (closed) subgroup of G.

(b) If H is open, then H is closed.

(c) If H contains a neighborhood of e, then H is open.

(d) H is open iff G/H (given the quotient topology) is a discrete space.

Next consider the separation properties T_1 (every point is closed) and T_2 (= Hausdorff). Recall that a topological space X is T_2 iff the diagonal $\Delta = \{(x,x) \in X \times X\}$ is closed.

PROPOSITION. A topological group G is T_2 iff G is T_1 iff $\{e\}$ is closed iff the intersection of all neighborhoods of e is $\{e\}$.

Take for example $SL(n,\mathbb{Q})$ or $SL(n,\mathbb{Z})$ with one of the subgroup topologies defined in 16.2. Since $T_c \subset T_a$, to show that each topology is T_2 it will sufficc to look at T_c. So it is enough to show that the intersection of all principal congruence subgroups is trivial; but this follows immediately from the fact that $\Gamma_p \cap \Gamma_q = \Gamma_{\mathrm{lcm}(p,q)}$.

Now consider connectedness properties. A space X is called totally disconnected if each of its components contains only a single point. (The component of $x \in X$ is the union of all connected subspaces of X containing x; the components are closed and connected, and partition X.) Note that a totally disconnected space is automatically T_1. Examples: $\mathbb{Z}$ or $\mathbb{Q}$ in the p-adic topology; $\mathbb{Q}$ in the usual topology of $\mathbb{R}$.

PROPOSITION. Let G be a topological group.

(a) If G is connected, it has no proper open subgroup, and is moreover generated by any open neighborhood of e.

(b) If H is the component of e in G, then H is a closed normal subgroup of G, whose cosets are the components of G. Moreover, G/H is totally disconnected.

(c) A product of totally disconnected groups is totally disconnected.

Finally, we discuss compactness properties. A space X is compact (not necessarily T_2) if each open covering of X contains a finite subcover. X is locally compact if each point has a compact neighborhood. It is not hard to see that if X is a compact T_2 space, the component of $x \in X$ is the intersection of all open-closed neighborhoods of x.

PROPOSITION. Let G, G_i $(i \in I)$ be topological groups, H a subgroup of G.

(a) G is locally compact iff e has a compact neighborhood.

(b) If G is locally compact and H is closed in G, then H is locally compact.

(c) If G is locally compact, so is the space G/H.

(d) $\prod G_i$ is locally compact iff all G_i are locally compact and almost all G_i are compact.

16.4 Profinite groups

In the arithmetic and congruence topologies on $\Gamma = SL(n,\mathbb{Z})$, a fundamental system of open neighborhoods of 1 consists in each case of a family of subgroups of finite index. So there is a natural process of "profinite completion", to be described below, leading to an exact sequence: $1 \to C \to \hat{\Gamma} \to \overline{\Gamma} \to 1$. This will yield a reformulation of the Congruence Subgroup Problem: Is C trivial?

Recall the definitions: I denotes a directed set, a partially ordered set in which each pair of indices has a common upper bound. An inverse system (in some category) consists of a collection of objects A_i $(i \in I)$ together with morphisms $f_{ji}: A_j \to A_i$ $(i \leq j)$ satisfying the coherence conditions: f_{ii} = identity; $f_{ji} f_{kj} = f_{ki}$ for $i \leq j \leq k$. For example, take the groups $\mathbb{Z}/p^i\mathbb{Z}$, along with the canonical maps $\mathbb{Z}/p^j\mathbb{Z} \to \mathbb{Z}/p^i\mathbb{Z}$ for $i \leq j$. More generally, take a collection $\{N_i, i \in I\}$ of normal subgroups of a group G, closed under finite intersections. Let $A_i = G/N_i$, and let the morphism $A_j \to A_i$ be the natural map (where $i \leq j$ in I means that $N_j \subset N_i$).

Given an inverse system as above, the inverse limit (or projective limit), denoted $\varprojlim A_i$ is defined to be the subgroup A of $\prod A_i$ consisting of those $(a_i)_{i \in I}$ for which $f_{kj}(a_k) = a_j$ whenever $j \leq k$. There is a universal property characterizing the inverse limit: Given an object B and morphisms $\tau_i: B \to A_i$ for which $\tau_j = f_{kj} \tau_k$ whenever defined, there exists a unique morphism $\tau: B \to A$

such that all triangles commute (σ_i = natural projection):

$$\begin{array}{ccc} B & \xrightarrow{\tau} & A \\ & \underset{\tau_i}{\searrow} & \downarrow \sigma_i \\ & & A_i \end{array}$$

A familiar example of this construction is of course $\mathbb{Z}_p = \varprojlim \mathbb{Z}/p^i\mathbb{Z}$.

In case our inverse system consists of finite groups, each given the discrete topology, the inverse limit has a topological group structure; it is in fact a closed subgroup of the topological product. We call such a group *profinite*. There is a nice characterization:

THEOREM. *A topological group G is profinite iff G is compact and totally disconnected.* (It is automatic that G is T_1, hence T_2.)

The idea of the proof is quite clear in one direction: If $G = \varprojlim G_i$, then $\prod G_i$ is compact (Tychonoff), while G is a closed (hence compact) subgroup. Moreover, G inherits total disconnectedness from $\prod G_i$. In the other direction, one observes that since G is compact T_2, the component of e, namely $\{e\}$, is the intersection of all open-closed neighborhoods of e. But each such neighborhood can be shown to contain an open normal subgroup N of G, necessarily of finite index since G is compact and G/N discrete. Now the universal property affords a map from G to the inverse limit of the finite groups G/N; this map can be seen to be an isomorphism of topological groups. (Moreover, the open normal subgroups N as above form a fundamental system of open neighborhoods of e in G.)

From this theorem flow a number of useful consequences:

PROPOSITION. *Let G, G_i $(i \in I)$ be topological groups.*

(a) *If G is profinite, so is every closed subgroup of G.*

(b) *If all G_i are profinite, so is $\prod G_i$.*

(c) *If G is profinite, H a closed normal subgroup of G, then G/H is profinite.*

(d) *If H is a normal subgroup of G, with both H and G/H*

profinite, then G is profinite.

Suppose now that we start with an arbitrary group G, together with a collection S of normal subgroups of finite index defining a subgroup topology on G. The quotients G/H $(H \in S)$ form an inverse system, so we may form $\tilde{G} = \varprojlim G/H$ and map G canonically into $\tilde{G}$ by sending g to the element of $\prod G/H$ $(H \in S)$ whose H-coordinate is gH. It is easy to check that the image of G is dense in $\tilde{G}$. Moreover, the map is injective provided the intersection of all groups in S is trivial. We call $\tilde{G}$ the profinite completion of G (relative to S).

Returning to $\Gamma = SL(n,\mathbb{Z})$, we see that the subgroup topologies T_a, T_c yield respective profinite completions $\hat{\Gamma}$, $\overline{\Gamma}$, where Γ may be viewed as a dense subgroup of each. Furthermore, the identity map from Γ (with topology T_a) to Γ (with topology T_c) is continuous, and can be extended (using the universal property of inverse limits) to a map $\hat{\Gamma} \to \overline{\Gamma}$. Since $\hat{\Gamma}$ is compact, while Γ is dense in $\overline{\Gamma}$, this map is surjective, leading to the promised exact sequence:

$$1 \to C \to \hat{\Gamma} \to \overline{\Gamma} \to 1 .$$

Of course, the kernel C is also profinite.

16.5 Completions of topological groups

For reasons which will become clear later, we are not content to study $SL(n,\mathbb{Z})$ alone; so we must look further for suitable "completions" of $SL(n,\mathbb{Q})$ relative to T_a, T_c. What is needed is an analogue of the procedure used to complete a metric space. In place of a metric and the resulting Cauchy sequences, we can use a "uniform structure" and the resulting Cauchy filters. The following brief outline is based on the development in Bourbaki [1, Chapters II, III]; though fairly elementary, the full development involves many details. (Cf. also Page [1].)

A uniform structure on a set X consists of a family E of subsets of $X \times X$ (called <u>entourages</u>) satisfying:

(U1) $V \in E$, $U \supset V \Rightarrow U \in E$

(U2) $U, V \in E \Rightarrow U \cap V \in E$

(U3) $V \in E \Rightarrow V \supset \Delta$ (the diagonal in X)

(U4) $V \in E \Rightarrow V^{-1} \in E$, where $V^{-1} = \{(y,x) \mid (x,y) \in V\}$

(U5) $V \in E \Rightarrow$ there exists $W \in E$ such that $W \circ W \subset V$, where $W \circ W = \{(x,z) \mid$ for some y, $(x,y), (y,z) \in W\}$.

When $(x,y) \in V \in E$, we say x is V-<u>close</u> to y. The idea is that entourages should provide a way to measure the nearness of a pair of points in X.

Given a uniform structure on X, we get a topology on X free of charge. This is most conveniently described by specifying the system of neighborhoods $N(x)$ for each $x \in X$: take all sets $V(x) = \{y \in X \mid (x,y) \in V\}$, where V runs over E. There is a simple criterion for X to be T_2: the intersection of all entourages must be Δ. We also get a notion of <u>uniform</u> <u>continuity</u> of a map $f: X \to Y$ between uniform spaces: given an entourage V of Y, there exists an entourage W of X such that whenever $(x,x') \in W$, we have $(f(x), f(x')) \in V$.

A subset F of E is called a <u>fundamental</u> <u>system</u> of entourages if each $V \in E$ includes some $W \in F$. Given F, we recover E (in view of (U1)) as the set of all subsets of $X \times X$ including some member of F. The axioms for a fundamental system are as follows:

(V1) $U, V \in F \Rightarrow$ there exists $W \in F$ such that $W \subset U \cap V$

(V2) $V \in F \Rightarrow \Delta \subset V$

(V3) $V \in F \Rightarrow$ there exists $W \in F$ such that $W \subset V^{-1}$

(V4) $V \in F \Rightarrow$ there exists $W \in F$ such that $W \circ W \subset V$.

The only examples of interest to us arise for a topological group G. As V runs over the neighborhoods of e in G, we let E_r be the family of all sets $V_r = \{(x,y) \in G \times G \mid yx^{-1} \in V\}$. Similarly,

$\mathcal{E}_l$ consists of all sets $V_l = \{(x,y) \mid x^{-1}y \in V\}$. It is easy to check that each of these gives a system of entourages for a uniform structure ("right" or "left") on G; moreover, the resulting topology coincides with the given topology on G. (Exercise: $\mathcal{E}_r$ yields the only uniform structure on G compatible with the given topology and such that its entourages are stable under all right translations.) When the group in question is $\mathbf{Z}$, with its p-adic topology, a fundamental system of entourages just consists of the sets $\{(x,y): x \equiv y \pmod{p^n}\}$ for all $n \geq 0$.

PROPOSITION. For a topological group G, the following are equivalent:

(a) $\mathcal{E}_r = \mathcal{E}_l$ (i.e., each V_r contains some W_l, and vice versa).

(b) For each neighborhood V of e, there is a neighborhood W of e such that $xWx^{-1} \subset V$ for all $x \in G$.

(c) The map $x \mapsto x^{-1}$ of $G \to G$ is uniformly continuous relative to $\mathcal{E}_r$

This criterion applies notably in the case when G has a fundamental system of neighborhoods of e consisting of normal subgroups. In particular, for either T_a or T_c, we need not distinguish between left and right uniform structures on $SL(n,Q)$.

To discuss completions, we use the notion of Cauchy filter. Recall that a filter $\mathcal{F}$ on a topological space X is a nonempty collection of nonempty subsets of X such that:

(1) $F \in \mathcal{F}$, $E \supset F \Rightarrow E \in \mathcal{F}$

(2) $E,F \in \mathcal{F} \Rightarrow E \cap F \in \mathcal{F}$.

A basis for $\mathcal{F}$ is a subset $\mathcal{B}$ of $\mathcal{F}$ such that each element of $\mathcal{F}$ includes some element of $\mathcal{B}$. (To be a basis of a filter, $\mathcal{B}$ must be a nonempty collection of nonempty subsets of X, with the property: $B,B' \in \mathcal{B} \Rightarrow$ there exists $B'' \in \mathcal{B}$ such that $B'' \subset B \cap B'$.) For example, the set $N(x)$ of neighborhoods of a point x is a filter, with basis given by a fundamental system of neighborhoods of x.

Call $x \in X$ a limit point of the filter F if $N(x) \subset F$. When X is T_2, such a limit point is unique and is just called the limit of F. In case X is a uniform space, with family E of entourages, we call a filter F a Cauchy filter if for any $V \in E$, there is a set $A \in F$ which is "V-small" (i.e., $A \times A \subset V$). Note that a filter with a limit x is certainly Cauchy. Note too that a uniformly continuous map takes a Cauchy filter basis to another such.

PROPOSITION. If X is a uniform space, F a Cauchy filter, then there exists a unique minimal Cauchy filter included in F (e.g., $N(x)$ in case F has a limit x).

Now call a uniform space X complete if every Cauchy filter has a limit.

PROPOSITION. Let X be a uniform space, Y a subspace.

(a) If X is locally compact (resp. discrete), then X is complete.

(b) If X is complete and Y is closed in X, then Y is complete (in the induced uniform structure).

(c) If X is T_2 and Y is complete (in the induced uniform structure), then Y must be closed in X.

(d) If Y is dense in X, and each Cauchy filter on Y has a limit in X, then X is complete.

Now we can state the main existence and uniqueness theorems.

EXTENSION THEOREM. Let X, Y be topological spaces (Y assumed regular), with A a dense subset of X and $f : A \to Y$ continuous. Assume the trace of $N(x)$ on A maps under f to a convergent filter basis, for all $x \in X$. Then f extends uniquely to a continuous map $X \to Y$.

COROLLARY. Let X, Y be uniform spaces, with Y complete and

T_2 (hence regular), A dense in X, $f:A \to Y$ uniformly continuous. Then f extends uniquely to a uniformly continuous map $X \to Y$.

COROLLARY. Let X, X' be complete, uniform T_2 spaces, with respective dense subspaces A, A'. Then any isomorphism (of uniform spaces) between A and A' extends uniquely to an isomorphism between X and X'.

EXISTENCE THEOREM. Let X be a T_2 uniform space. Then there exists a complete uniform T_2 space $\hat{X}$ and a uniformly continuous map $i:X \to \hat{X}$ (an isomorphism of X onto its image) such that: Given any uniformly continuous map $f:X \to Y$, where Y is a complete uniform T_2 space, there exists a unique uniformly continuous map $g:\hat{X} \to Y$ such that $g \circ i = f$.

To construct $\hat{X}$, one takes the set of all minimal Cauchy filters on X and defines a suitable uniform structure on it. Of course, i then maps $x \in X$ to its neighborhood filter.

If G is a topological group, call G complete if it is complete relative to both uniform structures $\mathcal{E}_r$, $\mathcal{E}_l$. (It suffices to check just one of them.) For either uniform structure it is easy to see that a continuous homomorphism $G \to H$ is automatically uniformly continuous.

EXTENSION THEOREM. Let G_1, G_2 be topological groups with respective dense subgroups H_1, H_2. Let G_2 be complete T_2, and let $f:H_1 \to H_2$ be a continuous homomorphism. Then f extends uniquely to a continuous homomorphism $G_1 \to G_2$, which is an isomorphism in case f is an isomorphism and G_1 is complete T_2.

EXISTENCE THEOREM. Let G be a T_2 topological group, for which $x \mapsto x^{-1}$ takes Cauchy filters to Cauchy filters. Then G is isomorphic to a dense subgroup of a complete T_2 group $\hat{G}$, which is unique up to isomorphism. Moreover, the closures in $\hat{G}$ of the neighborhoods of e in G form a fundamental system of neighborhoods of e .

It is worth glancing back at the discussion of profinite groups in 16.4. If we start with a group G, together with a collection S of normal subgroups of finite index closed under finite intersections and intersecting in e, we get a T_2 topology on G by taking S as fundamental system of neighborhoods of e. (We did this for $SL(n,\mathbb{Z})$ using T_a or T_c.) Then $\tilde{G} = \varprojlim G/N$ $(N \in S)$ is profinite, and G embeds naturally in $\tilde{G}$ as a dense subgroup. Since $\tilde{G}$ is compact (hence complete) and T_2, we conclude from the above uniqueness result that $\tilde{G}$ is the completion $\hat{G}$.

16.6 The congruence kernel

Thanks to 16.5, $G = SL(n,\mathbb{Q})$ and $\Gamma = SL(n,\mathbb{Z})$ admit completions with respect to the two subgroup topologies T_a, T_c. For T_a write $\hat{G},\hat{\Gamma}$, and for T_c write $\bar{G},\bar{\Gamma}$. We know also that $\hat{\Gamma},\bar{\Gamma}$ can be viewed as profinite completions in the sense of 16.4, and that there is an exact sequence: (*) $1 \to C \to \hat{\Gamma} \to \bar{\Gamma} \to 1$. Moreover, $\hat{\Gamma}$ and $\bar{\Gamma}$ may be identified with the respective closures of Γ in $\hat{G}$ and $\bar{G}$, using the uniqueness of completions.

We observed earlier that T_c induces on G the topology it receives as a subgroup of $G_{\mathbb{A}^f} = SL(n,\mathbb{A}^f)$. The strong approximation theorem 14.3 says that G is dense in $G_{\mathbb{A}^f}$. The latter group is complete, being locally compact, so the uniqueness of completions allows us to identify it with $\bar{G}$. Under this identification, $\bar{\Gamma}$ corresponds to $\prod_p SL(n,\mathbb{Z}_p)$.

The identity map from G (given the topology T_a) to G (given the topology T_c) is continuous. The extension theorem therefore furnishes a continuous map $\pi:\hat{G} \to \bar{G}$. Let us analyze the restriction of π to $\hat{\Gamma}$, whose kernel C we call the congruence kernel. We claim first that $\pi(\hat{G}) = \bar{G}$. As pointed out earlier, the closures in

$\overline{G}$ of the neighborhoods of 1 in G form a fundamental system of neighborhoods of 1. So $\overline{\Gamma}$ is an open subgroup of $\overline{G}$, and lies in $\pi(\hat{G})$. But a subgroup of a topological group containing an open set is itself open (hence closed), so $\pi(\hat{G})$ is closed in $\overline{G}$. It contains the dense subgroup G and therefore equals $\overline{G}$, as claimed.

Consider next Ker π. Say $x \in \hat{G}$, $\pi(x) = 1$. Then x is a minimal Cauchy filter on G converging to 1 in T_c, i.e., all principal congruence subgroups of Γ lie in this filter. But then the trace of this filter on Γ is a Cauchy filter on Γ (in the topology T_a) which converges to 1 in T_c. The filter on Γ must converge to a point of $\hat{\Gamma}$(= closure of Γ in $\hat{G}$), so we conclude that $x \in \hat{\Gamma}$ and therefore $x \in C$. This implies that $C = \text{Ker}\ \pi$.

We have now succeeded in restating the Congruence Subgroup Problem as a question about the exact sequence: (**) $1 \to C \to \hat{G} \to \overline{G} \to 1$. Is C trivial? This could of course be posed just for the sequence (*). But the groups $\hat{G}$, $\overline{G}$ are much more amenable to study from the viewpoint to be developed below.

It is worth emphasizing that the only special feature of SL(n,Q) used here is the strong approximation property. Apart from this, our formulation of the problem is quite general.

§17. <u>The congruence kernel of</u> SL(n,Z)

The main goal of this section is to show that when $G = SL(n,Q)$, $\Gamma = SL(n,Z)$, $n \geq 3$, the congruence kernel C defined in §16 lies in the <u>center</u> of both $\hat{G}$ and $\hat{\Gamma}$. In this way we are led to the study of central extensions. We follow the approach of Bass-Lazard-Serre [1], concentrating on Γ. For other groups and other rings of algebraic integers, the arguments used by Matsumoto [1] are analogous, but appreciably more intricate.

17.1 <u>Some consequences of the invariant factor theorem</u>

To get some information about the structure of Γ, we appeal to the invariant factor theorem. In one concrete version this states: Given a rectangular matrix A with integral entries, there exist invertible matrices P and Q over $\mathbb{Z}$ for which PAQ has "diagonal" form

$$\begin{pmatrix} d_1 & & \\ & d_2 & \\ & & \ddots \end{pmatrix}$$

with zeros off the "diagonal" and with each d_i dividing d_{i+1}; the d_i are then unique up to sign. This specializes to a familiar algorithm for diagonalization of an arbitrary element of Γ, using two kinds of matrix operations: addition of a multiple of one row (or column) to another, interchange of two rows (or columns) with possible change of signs. In fact the second type of operation just involves successive uses of the first type, as illustrated when $n=2$:

$\begin{pmatrix} 1 & 1 \\ 0 & 1 \end{pmatrix}\begin{pmatrix} 1 & 0 \\ -1 & 1 \end{pmatrix}\begin{pmatrix} 1 & 1 \\ 0 & 1 \end{pmatrix} = \begin{pmatrix} 0 & 1 \\ -1 & 0 \end{pmatrix}$. The first type of operation can be carried out by left or right multiplication by an <u>elementary matrix</u>, by which we mean one of the form $x_{ij}(t)$, $i \neq j$, the matrix differing from the identity matrix only by the presence of t in the (i,j) position.

<u>THEOREM 1</u>. (a) Γ <u>is generated by elementary matrices.</u>

(b) $\Gamma = (\Gamma,\Gamma)$ <u>provided</u> $n \geq 3$.

<u>Proof</u>. (a) We need only refine slightly the algorithm mentioned above. If $x \in \Gamma$, we have $e_r \ldots e_1 x f_1 \ldots f_s = \mathrm{diag}(d_1,\ldots,d_n)$, where the e_i and f_j are elementary. Since the determinant is 1, each $d_i = \pm 1$, with an even number of negative entries. A further product of elementary matrices will change any given pair of signs, as illustrated when $n = 2$ by the matrix equation above, together with

$\begin{pmatrix} 0 & 1 \\ -1 & 0 \end{pmatrix}^2 = \begin{pmatrix} -1 & 0 \\ 0 & -1 \end{pmatrix}$. So x is eventually a product of elementary matrices.

(b) Note that commutators in Γ behave as follows: $(x_{ij}(s),$

$x_{jk}(t)) = x_{ik}(st)$ for $i \neq k$, while $(x_{ij}(s), x_{k\ell}(t)) = 1$ if $j \neq k$ and $i \neq \ell$. Given $i \neq k$, find $j \neq i,k$ (possible since $n \geq 3$). Then $(x_{ij}(s), x_{jk}(1)) = x_{ik}(s) \in (\Gamma,\Gamma)$. But Γ is generated by matrices $x_{ik}(s)$ according to part (a), so $\Gamma = (\Gamma,\Gamma)$. □

There is a more abstract version of the invariant factor theorem: Given a free abelian group L of rank n, and a subgroup K (necessarily free abelian of rank $\leq n$), there exist ordered bases $(e_1,..,e_n)$ of L, $(f_1,...,f_r)$ of K, such that $f_i = d_i e_i$ for some $d_i \in \mathbb{Z}$ (where $d_i \mid d_{i+1}$, the d_i unique up to sign). Now call $a = (a_1,...,a_n) \in \mathbb{Z}^n$ <u>unimodular</u> if the a_i are relatively prime, i.e., generate the unit ideal in $\mathbb{Z}$.

<u>THEOREM 2</u>. Γ <u>acts transitively on the set of unimodular elements of</u> $\mathbb{Z}^n$ $(n \geq 2)$.

<u>Proof</u>. Let $a = (a_1,...,a_n) \in \mathbb{Z}^n$ be unimodular. It clearly suffices to show that a can be sent to the first standard basis vector $(1,0,...,0)$. Set $K = \mathbb{Z}a$, $L = \mathbb{Z}^n$, and use the invariant factor theorem to find bases $(e_1,...,e_n)$ of $\mathbb{Z}^n$, (f_1) of $\mathbb{Z}a$, for which $f_1 = de_1$. Evidently $f_1 = \pm a$, so $d \mid a_i$ for all i. Since the a_i are assumed to be relatively prime, we find $d = \pm 1$; we may as well assume that $e_1 = a$. Now let $g \in GL(n,\mathbb{Z})$ take the standard basis of $\mathbb{Z}^n$ to the basis $(e_1,...,e_n)$. The vector a occupies the first column of g, whose determinant is ± 1. If $\det g = -1$, just multiply the second column of g by -1 to get a matrix in Γ which sends a to $(1,0,...,0)$. □

17.2 <u>Congruence subgroups and</u> q-<u>elementary subgroups</u>

Until further notice, $\Gamma = SL(n,\mathbb{Z})$ with $n \geq 3$. We also make the convention that q denotes a positive integer.

Let E_q be the smallest normal subgroup of Γ containing all elementary matrices of the form $x_{ij}(qa)$, $a \in \mathbb{Z}$. Evidently $E_q \subset \Gamma_q$.

But it is not at all obvious that E_q must be of finite index in Γ; this will emerge shortly. First we make explicit the special role played by the q-elementary subgroups E_q in the structure theory of Γ.

<u>LEMMA</u>. <u>Any subgroup</u> H <u>of finite index in</u> Γ <u>includes some</u> E_q.

<u>Proof</u>. Without loss of generality we may assume that H is normal in Γ, since any subgroup of finite index includes a normal subgroup of finite index. Say $[\Gamma:H] = q$, so $x^q \in H$ for all $x \in \Gamma$. We claim that $E_q \subset H$, for which it is enough to observe that $x_{ij}(qa) \in H$, $a \in \mathbb{Z}$. But $x_{ij}(qa) = x_{ij}(a)^q$. □

We saw in 17.1 that Γ acts transitively on unimodular elements of $\mathbb{Z}^n$. The main theorem of this section asserts that E_q acts transitively on certain special sets of unimodular vectors. We shall postpone the proof until 17.4, in order to point out several important corollaries.

<u>THEOREM</u>. <u>Let</u> $a = (a_1,\ldots,a_n)$, $a' = (a_1',\ldots,a_n') \in \mathbb{Z}^n$ <u>be unimodular vectors, with</u> $a_i \equiv a_i' \pmod{q\mathbb{Z}}$ <u>for all</u> i. <u>Then there exists</u> $s \in E_q$ <u>such that</u> $s(a) = a'$.

For the first corollary, we have to let $n \geq 3$ vary, so we write $\Gamma(n) = SL(n,\mathbb{Z})$, with $\Gamma(n-1)$ embedded in $\Gamma(n)$ via $x \mapsto \begin{pmatrix} x & 0 \\ 0 & 1 \end{pmatrix}$. (Here $n-1 = 2$ is permitted.) Write $\Gamma_q(n)$, $E_q(n)$, etc.

<u>COROLLARY</u> 1. $\Gamma_q(n) = E_q(n)\,\Gamma_q(n-1)$.

<u>Proof</u>. Let $e_n = (0,\ldots,0,1)$. If $t \in \Gamma_q(n)$, then $t(e_n)$ is again unimodular and $t(e_n) \equiv e_n \pmod q$. Applying the theorem to the pair e_n, $t(e_n)$, we get $s \in E_q(n)$ for which $st(e_n) = e_n$. In block form, st looks like: $\begin{pmatrix} x & 0 \\ y & 1 \end{pmatrix}$. Here q divides all entries of y, since $st \in \Gamma_q(n)$, while $x \in \Gamma_q(n-1)$. Set $z = -yx^{-1} \in q\mathbb{Z}^{n-1}$. Then $\begin{pmatrix} x & 0 \\ 0 & 1 \end{pmatrix} = \begin{pmatrix} 1 & 0 \\ z & 1 \end{pmatrix}\begin{pmatrix} x & 0 \\ y & 1 \end{pmatrix} = rst \in \Gamma_q(n-1)$. Note that

r is a product of $n-1$ commuting elements of $E_q(n)$ and therefore lies in $E_q(n)$, so $rs \in E_q(n)$. This yields the desired factorization of t. □

<u>COROLLARY</u> 2. $(\Gamma,\Gamma_q) \subset E_q$.

<u>Proof</u>. Recall that we are writing $(a,b) = aba^{-1}b^{-1}$. We require two commutator identities:

(1) $(ab,c) = (b,c)^{a^{-1}}(a,c)$, where $(b,c)^a = a^{-1}(b,c)a$,

(2) $(a,bc) = (a,b)\ (a,c)^{b^{-1}}$.

Now let $s \in \Gamma$, $t \in \Gamma_q$. Suppose we already know that $(s,t) \in E_q$ in case s is an elementary matrix. Writing an arbitrary s as a product $s_1 \ldots s_m$ of elementary matrices (see 17.1), we can use (1) repeatedly to get $(s,t) \in E_q$, since $E_q \lhd \Gamma$. So it is enough to prove that $(s,t) \in E_q$ when s is <u>elementary</u>.

We claim that s is conjugate in Γ to a matrix of the special form $r = \left(\begin{array}{c|c} 1 & 0 \\ \hline x & 1 \end{array}\right)$. To see this, write $w_{ij} = x_{ij}(1)\ x_{ji}(-1)x_{ij}(1)$. Up to change of sign, w_{ij} just permutes the ith and j^{th} rows of the identity matrix. Conjugating an elementary matrix s by suitable w_{ij} will clearly result in a matrix of the special form r.

Now suppose we know that $(r,u) \in E_q$ whenever r has the above form and $u \in \Gamma_q$. It will follow that $(s,t) \in E_q$: write $s = r^g$ $(g \in \Gamma)$, and $t = u^g$ (using the fact that $\Gamma_q \lhd \Gamma$), so $(s,t) = (r^g,u^g) = (r,u)^g \in E_q$ (since $E_q \lhd \Gamma$) .

Finally, we can treat this special case (r,u) via Corollary 1. Write $u = vw$, where $v \in E_q$, $w \in \Gamma_q(n-1)$. Identity (2) implies that $(r,u) = (r,v)(r,w)^{v^{-1}}$, with $(r,v) \in E_q$. We have to show that $(r,w) \in E_q$, where $w = \left(\begin{array}{c|c} a & 0 \\ \hline 0 & 1 \end{array}\right)$, $a \in \Gamma_q(n-1)$. Direct computation yields $(r,w) = \left(\begin{array}{c|c} 1 & 0 \\ \hline x-xa^{-1} & 1 \end{array}\right)$. But $x-xa^{-1} \equiv 0 \pmod q$, forcing $(r,w) \in E_q$. □

<u>COROLLARY</u> 3. $[\Gamma:E_q] < \infty$.

Proof. Recall from 17.1 that $\Gamma = (\Gamma,\Gamma)$. So we have the following inclusions, thanks to Corollary 2:

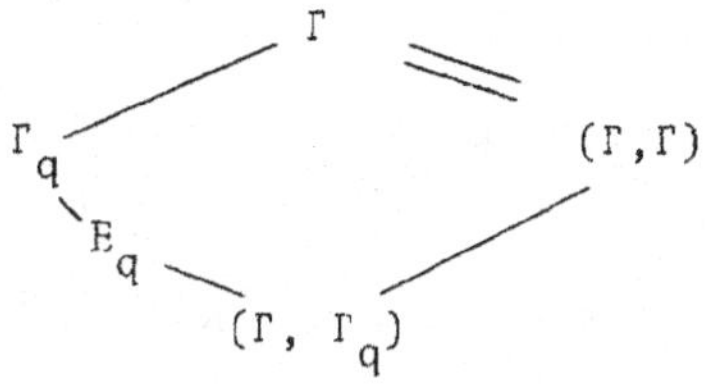

It will suffice to show that (Γ, Γ_q) has finite index in Γ. But this is an immediate consequence of a standard lemma, recalled in 17.3 below. □

17.3 A finiteness lemma

LEMMA. *Let* A *be a group,* B *a normal subgroup. If* A/B *and* $A/(A,A)$ *are finite, then* $A/(A,B)$ *is finite.*

Proof. Consider the diagram:

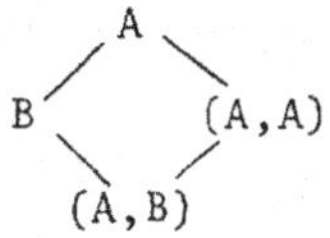

Set $\bar{A} = A/(A,B)$, so the canonical images of the indicated groups are $\bar{B}$, $\overline{(A,A)} = (\bar{A},\bar{A})$, $\overline{(A,B)} = (\bar{A},\bar{B}) = 1$. By hypothesis, $\bar{A}/\bar{B}$ and $\bar{A}/(\bar{A},\bar{A})$ are finite. Note also that $\bar{B} \subset Z(\bar{A})$, forcing $\bar{A}/Z(\bar{A})$ to be finite. It is therefore enough to prove the following statement: *If* A *is a group for which* $A/(A,A)$ *and* $A/Z(A)$ *are finite, then* A *itself is finite.*

Set $S = \{(a,b) \mid a,b \in A\}$. If $[A:Z(A)] = m$, with coset representatives $a_1,\dots,a_m$, we claim that Card $S \le m^2$. To see this, let $(a,b) \in S$, with $a \in a_iZ(A)$ and $b \in a_jZ(A)$. A quick computation shows that $(a,b) = (a_i,a_j)$. So there are no more than m^2 distinct commutators.

Now adopt some fixed ordering of S. Conjugation allows us to reorder a product of commutators without changing the length of the product, so an arbitrary $x \in (A,A)$ can be written as $\prod_{s\in S} s^{m(s)}$ in the given ordering. To show that (A,A) and hence A itself is finite, it will suffice to show that each s^{m+1} is a product of at most m commutators (so each element of (A,A) is a product of at most m^3 factors from S). Write $s = (a,b) \in S$, so $s^{m+1} = (a,b)^m \cdot (a,b) =$ [since $(a,b)^m$ is central in A] $b^{-1}(a,b)^m b(a,b) = b^{-1} \cdot (a,b)^{m-1}(a,b^2)b = ((a,b)^b)^{m-1}(a,b^2)^b$. □

(This proof follows Humphreys [1, Lemma 17.1A].)

17.4 Proof of the theorem

We will require two technical lemmas.

LEMMA 1. *Let $R = \mathbb{Z}/t\mathbb{Z}$, $t \neq 0$, and suppose $a = (a_1,\ldots,a_m) \in R^m$ is unimodular. Then there exist $b_2,\ldots,b_m \in R$ such that $a_1 + b_2a_2 + \ldots + b_ma_m$ is a unit.*

Proof. (a) Suppose R is a field (i.e., t is a prime). Then "unit" means nonzero and "unimodular" means that not all a_i are 0. If $a_1 \neq 0$, set all $b_i = 0$. If $a_1 = 0$ but $a_k \neq 0$, set $b_k = 1$, all other $b_i = 0$.

(b) Next suppose R is a direct sum of fields (i.e., t is a product of distinct primes), say $R = F_1 \oplus\ldots\oplus F_r$. Write $c \in R$ as $c^{(1)}\oplus\ldots\oplus c^{(r)}$. Now c is a unit provided all $c^{(k)}$ are nonzero, while for $a \in R^m$ to be unimodular means that for each index k, not all of $a_1^{(k)},\ldots,a_m^{(k)}$ are zero. Applying (a) to each field F_k, we find $b_2^{(k)},\ldots,b_m^{(k)}$ such that $a_1^{(k)} + b_2^{(k)}a_2^{(k)}+\ldots+ b_m^{(k)}a_m^{(k)} \neq 0$ in F_k. Then set $b_i = b_i^{(1)} \oplus \ldots \oplus b_i^{(r)}$.

(c) In general, let J be the Jacobson radical (= nilradical) of R, so $\bar{R} = R/J$ is a direct sum of fields. Since $a \in R^m$ is unimodular, $\bar{a} = (\bar{a}_1,\ldots,\bar{a}_m) \in \bar{R}^m$ is clearly unimodular. So applying (b) to $\bar{R}$, we find $\bar{b}_2,\ldots,\bar{b}_m \in \bar{R}$ for which $\bar{a}_1 + \bar{b}_2\bar{a}_2+\ldots+\bar{b}_m\bar{a}_m$ is a unit. Thus $c = a_1 + b_2a_2+\ldots+b_ma_m$ is a unit (mod J): there exists $d \in R$, $n \in J$ so that $cd = 1 + n$. But n is nilpotent, so $1 + n$ is a unit, forcing c to be a unit in R. □

LEMMA 2. Let $a,b \in \mathbf{Z}^n$. Suppose $I \subset \{1,2,\ldots,n\}$ satisfies : $a_i = b_i$ $(i \in I)$, and for $j \notin I$, $b_j \equiv a_j \pmod{qq'}$, where $q \geq 1$ and where $q'\mathbf{Z}$ is the ideal generated by all a_i, $i \in I$. Then there exists $s \in E_q$ for which $s(a) = b$.

Proof. If $j \notin I$, we have by hypothesis an equation $b_j = a_j + q \sum_{i \in I} t_{ij} a_i$. Define $s = \prod x_{ji}(qt_{ij})$, product over $j \notin I, i \in I$, where the order is immaterial since the elementary matrices involved all commute. Evidently $s \in E_q$, and direct computation shows that $s(a) = b$. This is best illustrated by an example: say $n = 3$, and $b_1 = a_1$, $b_2 = a_2$, $b_3 = a_3 + qt_{13}a_1 + qt_{23}a_2$; then $s = x_{31}(qt_{13})x_{32}(qt_{23})$, with

$$\begin{pmatrix} 1 & 0 & 0 \\ 0 & 1 & 0 \\ qt_{13} & 0 & 1 \end{pmatrix}\begin{pmatrix} 1 & 0 & 0 \\ 0 & 1 & 0 \\ 0 & qt_{23} & 1 \end{pmatrix}\begin{pmatrix} a_1 \\ a_2 \\ a_3 \end{pmatrix} = \begin{pmatrix} 1 & 0 & 0 \\ 0 & 1 & 0 \\ qt_{13} & 0 & 1 \end{pmatrix}\begin{pmatrix} a_1 \\ a_2 \\ qt_{23}a_2 + a_3 \end{pmatrix} = \begin{pmatrix} a_1 \\ a_2 \\ qt_{13}a_1 + qt_{23}a_2 + a_3 \end{pmatrix} = \begin{pmatrix} b_1 \\ b_2 \\ b_3 \end{pmatrix}.$$

Now we can prove the theorem. We are given unimodular vectors $a \equiv a' \pmod q$, and we must find $s \in E_q$ such that $s(a) = a'$. Of course we may assume that $q > 1$, since $E_1 = \Gamma$ acts transitively on unimodular vectors (17.1).

We claim it will suffice to treat the case $a' = e_1 = (1,0,\ldots,0)$. Indeed, assuming this case done, we can find $t \in \Gamma$ sending a' to e_1, then $s \in E_q$ sending $t(a)$ to e_1, so that $t^{-1}st \in E_q$ sends a to a' as desired.

To send a to e_1, we will use the lemmas above and proceed in several steps, indicated schematically as follows:

	$(a_1,a_2,\ldots,a_n)$		
step 1	$\downarrow$	$s_1 \in E_q$	$I = \{3,4,\ldots,n\}$
	$(a_1,b,a_3,\ldots,a_n)$		
step 2	$\downarrow$	$s_2 \in E_q$	$I = \{1,2\}$
	$(a_1,b,r,0,\ldots,0)$		

step 3 $\downarrow$ $t = x_{13}(1) \in \Gamma$

$(1,b,r,0,\ldots,0)$

step 4 $\downarrow$ $s_3 \in E_q$ $I = \{1\}$

$(1,0,\ldots,0)$

step 5 $\downarrow$ t^{-1}

$(1,0,\ldots,0)$

We have to explain these steps. We are given $a_1 \equiv 1 \pmod q$, say $a_1 = 1-r$ ($r \in q\mathbf{Z}$); all other $a_i \equiv 0 \pmod q$. Now a is unimodular, so $1 = c_1a_1 + \ldots + c_na_n$ for some c_i. Thus (reducing mod a_1): $\bar{1} = \bar{c}_2\bar{a}_2 + \ldots + \bar{c}_n\bar{a}_n = \bar{c}_2\bar{a}_2 + \bar{c}_3\overline{ra}_3 + \ldots + \bar{c}_n\overline{ra}_n$, since $\bar{r} = \bar{1}$. So $(a_2, ra_3,\ldots,ra_n)$ is unimodular (mod a_1). By Lemma 1, there exist $t_2,\ldots, t_n$ such that $b = a_2 + \sum_{i>2} rt_ia_i$ is a unit (mod a_1), i.e., $bk = 1+\ell a_1$ for some k,ℓ. In particular, a_1 and b are relatively prime. Now it remains to justify the application of Lemma 2 at steps 1,2,4 above.

Step 1: Here $I = \{3,\ldots,n\}$, $b = a_2 + \sum_{i>2} rt_ia_i$. We know that $q|r$, while q' divides all a_i $(i > 2)$, so qq' divides each term of the sum. From Lemma 2 we get $s_1 \in E_q$ as desired.

Step 2: Here $I = \{1,2\}$. Since a_1 and b are relatively prime, $q' = 1$, and we require only that $a_i \equiv 0 \pmod q$ for $i \geq 4$, $a_3 \equiv r \pmod q$, both of which are clear.

Step 3: Since $a_1 + r = 1$ (by choice of r), $t = x_{13}(1)$ clearly has the desired effect.

Step 4: Here $I = \{1\}$, with $q' = 1$. So we require only that q divide b and r, which is evident.

Combining steps 1-5, we obtain $s = (t^{-1}s_3t)s_2s_1 \in E_q$, taking a to e_1. □

17.5 The congruence kernel

Corollary 2 in 17.2 states that $(\Gamma,\Gamma_q) \subset E_q$. In other words, Γ_q/E_q lies in the center of Γ/E_q. It follows immediately that $\varprojlim \Gamma_q/E_q$ is central in $\varprojlim \Gamma/E_q$.

According to Corollary 3, Γ/E_q is finite. Therefore $\varprojlim \Gamma/E_q$ is profinite, hence compact T_2, hence complete. But by Lemma 17.2, Γ is dense in this inverse limit, so the Extension Theorem 16.5 implies that $\hat{\Gamma} \cong \varprojlim \Gamma/E_q$. We also know that $\overline{\Gamma} \cong \varprojlim \Gamma/\Gamma_q$. Moreover, the universal property of inverse limits yields a natural map from $\varprojlim \Gamma/E_q$ to $\varprojlim \Gamma/\Gamma_q$, which is clearly the identity map on the respective dense subgroups identified with Γ and which has kernel equal to $\varprojlim \Gamma_q/E_q$. It follows from these considerations that the sequence: $1 \to C \to \hat{\Gamma} \to \overline{\Gamma} \to 1$, may be identified with the sequence:

$1 \to \varprojlim \Gamma_q/E_q \to \varprojlim \Gamma/E_q \to \varprojlim \Gamma/\Gamma_q \to 1$. As a result, $C \cong \varprojlim \Gamma_q/E_q$ is <u>central</u> in $\hat{\Gamma}$. Consider in turn the sequence: $1 \to C \to \hat{G} \xrightarrow{\pi} \overline{G} \to 1$.

<u>THEOREM</u>. C <u>is central in</u> $\hat{G}$.

<u>Proof</u>. Let $\hat{H} = C_{\hat{G}}(C)$ be the centralizer of C in $\hat{G}$, $H = \hat{H} \cap G$. Since $C \lhd \hat{G}$, $\hat{H} \lhd \hat{G}$ and $H \lhd G$. We claim that $H = G$. Since $\Gamma \subset H$ by the preceding discussion, this is an obvious consequence of the well known fact that G has no proper normal subgroups outside its (finite) center. Alternatively, we can show directly that all elementary matrices (which generate G) lie in H: conjugating $x_{ij}(1) \in \Gamma$ by a suitable diagonal matrix in G of determinant 1, say $\mathrm{diag}(t_1, \ldots, t_n)$ where $t_i = t$, $t_j = 1$, yields $x_{ij}(t) \in H$. Now G centralizes C, so for fixed $c \in C$ the continuous map $g \mapsto (c,g)$ from $\hat{G}$ to $\hat{G}$ has image 1 on the dense subgroup G, hence is the trivial map. So $\hat{G}$ centralizes C. □

<u>Remark</u>. If we let n vary as in 17.2, we have a sequence of congruence kernels $C(n) \cong \varprojlim \Gamma_q(n)/E_q(n)$. From Corollary 1 of 17.2 we obtain surjective homomorphisms $C(3) \to C(4) \to C(5) \to \ldots$ So if we knew that $C(3)$ is trivial, the Congruence Subgroup Problem would be settled affirmatively for all $n \geq 3$.

17.6 *Universal property of the extension*

If $1 \to X \to Y \to Z \to 1$ is an exact sequence of groups, with X central in Y, we call the sequence (or just the group Y) a *central extension* of Z.

THEOREM. *Let* $1 \to F \to E \xrightarrow{p} \bar{G} \to 1$ *be a central extension of topological groups (i.e.,* F *maps isomorphically onto a subgroup of* E, *and* p *identifies* $\bar{G}$ *with the topological quotient* E/F). *Assume that* F *is profinite, and that* E *has a subgroup mapped isomorphically by* p *onto* G *(the extension is "trivial with respect to* G"). *Then there is a continuous homomorphism* $\phi\colon \hat{G} \to E$ *making the following diagram commute:*

$$\begin{array}{ccccccccc} 1 & \to & C & \to & \hat{G} & \xrightarrow{\pi} & \bar{G} & \to & 1 \\ & & & & \downarrow\phi & & \| & & \\ 1 & \to & F & \to & E & \to & \bar{G} & \to & 1 \end{array}$$

Proof. Observe that E must be locally compact. The proof of the theorem goes in two steps. First we restrict attention to the situation:

$$\begin{array}{ccccccccc} 1 & \to & C & \to & \hat{\Gamma} & \to & \bar{\Gamma} & \to & 1 \\ & & & & & & \| & & \\ 1 & \to & F' & \to & L & \to & \bar{\Gamma} & \to & 1 \end{array}$$

Here $L = p^{-1}(\bar{\Gamma})$. Set-theoretically, we can just define ϕ on $\Gamma \subset \hat{\Gamma}$ so that the resulting square commutes:

$$\begin{array}{ccc} \Gamma & \xrightarrow{\pi} & \Gamma \\ \phi\downarrow & & \| \\ \Gamma & \xrightarrow[p]{} & \Gamma \end{array}$$

(We have assumed that $G \subset E$, so $\Gamma \subset L$.) It has to be checked that ϕ is continuous. Observe that L is profinite (16.4), as well as open in E due to the continuity of p. Thus the topology induced on the copy of Γ in L has a fundamental system of neighborhoods of 1 consisting of certain subgroups of finite index; so their inverse images under ϕ are open. Because ϕ is continuous, it extends to a continuous homomorphism $\phi : \hat{\Gamma} \to L$.

Now $G \cap \overline{\Gamma} = \Gamma$, whence $G \cap L = \Gamma$ (in E). So we can define $\phi : G \to G$ unambiguously as the identity map. Since ϕ is continuous on the open subset Γ, it is continuous on G. As a locally compact group, E is complete; so ϕ extends to a continuous map $\hat{G} \to E$. Since G is dense in $\hat{G}$, the resulting diagram commutes. □

In §18 and §19 we shall focus on the central extensions of SL(n,K) where K is for example a local field $\mathbb{Q}_p$; then in §20 we shall consider how to fit together such extensions to get a suitable universal central extension of $\overline{G}$ (trivial with respect to G) as described in the theorem. The local data involved will then determine the congruence kernel C.

§18. The Steinberg group

The object of this section is to construct a universal central extension of SL(n,K) when K is a field and $n \geq 3$. The construction is due to Steinberg, for arbitrary Chevalley groups (cf. Steinberg [1], [2,§6-7] and Milnor [1, §5,8,9]).

18.1 Generators and relations

Start with any field K, letting $G = SL(n,K)$ for $n \geq 3$. As in §17, $x_{ij}(t)$ ($i \neq j$, $t \in K$) denotes the elementary matrix with t in the (i,j) position. These matrices generate G. If $u \in K^*$, we set $w_{ij}(u) = x_{ij}(u)\, x_{ji}(-u^{-1})\, x_{ij}(u)$, $h_{ij}(u) = w_{ij}(u)\, w_{ij}(-1)$. These are monomial matrices, e.g, $h_{ij}(u)$ is diagonal with i^{th} entry u, j^{th} entry u^{-1}, other entries 1.

The generators $x_{ij}(t)$ satisfy the following relations:

(R1) $x_{ij}(s)\, x_{ij}(t) = x_{ij}(s+t)$

(R2) $(x_{ij}(s), x_{j\ell}(t)) = x_{i\ell}(st)$ if $i \neq \ell$,

$(x_{ij}(s), x_{k\ell}(t)) = 1$ if $j \neq k$, $i \neq \ell$.

(For other pairs of indices, commutators are much more complicated to describe. Notice that (R2) is vacuous for SL(2,K), which is why we require $n \geq 3$ from the outset. For $n = 2$, Steinberg introduces another relation.)

Now define $\tilde{G} = St(n,K)$ to be the abstract group with generators $\tilde{x}_{ij}(t)$ $(i \neq j,\ t \in K)$, subject only to the relations corresponding to (R1), (R2). Let $\phi: \tilde{G} \to G$ be the canonical epimorphism sending $\tilde{x}_{ij}(t)$ to $x_{ij}(t)$. Our goal is to prove that $1 \to \mathrm{Ker}\phi \to \tilde{G} \to G \to 1$ is a central extension having a suitable universal property. To this end we look closely at certain subgroups of $\tilde{G}$ and try to imitate the Bruhat decomposition in G, cf. §12.

Remark. St(n,R) could be defined just as well for an arbitrary ring R with 1, but then the determination of $\mathrm{Ker}\phi$ becomes much harder; this amounts to studying the K_2 functor of Milnor in algebraic K-theory.

18.2 The upper unitriangular group

Let U (resp. $\tilde{U}$) be the subgroup of G (resp. $\tilde{G}$) generated by those $x_{ij}(t)$ (resp. $\tilde{x}_{ij}(t)$) for which $i < j$. If $X_{ij} = \{x_{ij}(t)\}$ and $\tilde{X}_{ij} = \{\tilde{x}_{ij}(t)\}$, relation (R1) makes it plain that ϕ maps $\tilde{X}_{ij}$ isomorphically onto X_{ij}.

PROPOSITION. *ϕ maps $\tilde{U}$ isomorphically onto U.*

Proof. It first has to be observed that each element of U is expressible uniquely as a product of elements from the various 1-parameter groups X_{ij}, taken (say) in lexicographic order. For example, when $n = 3$:

$$\begin{pmatrix} 1 & a & c \\ 0 & 1 & b \\ 0 & 0 & 1 \end{pmatrix} = \begin{pmatrix} 1 & a & 0 \\ 0 & 1 & 0 \\ 0 & 0 & 1 \end{pmatrix} \begin{pmatrix} 1 & 0 & c-ab \\ 0 & 1 & 0 \\ 0 & 0 & 1 \end{pmatrix} \begin{pmatrix} 1 & 0 & 0 \\ 0 & 1 & b \\ 0 & 0 & 1 \end{pmatrix} .$$

In general, the idea is to begin with an arbitrary product of elements $x_{ij}(t)$ in U, then rewrite in the desired order by repeated use of relations (R1) and (R2). A typical step (for $n = 3$) would replace

$x_{23}(s)x_{12}(t)$ by $x_{12}(t)x_{23}(s)x_{13}(-st) = x_{12}(t)x_{13}(-st)x_{23}(s)$. Since this factorization in U depends only on (R1) and (R2), the same thing can be done in $\tilde{U}$; so the proposition follows. □

18.3 The monomial group

Define $\tilde{w}_{ij}(u) = \tilde{x}_{ij}(u)\, \tilde{x}_{ji}(-u^{-1})\, \tilde{x}_{ij}(u)$ and $\tilde{h}_{ij}(u) = \tilde{w}_{ij}(u) \cdot \tilde{w}_{ij}(-1)$ for $i \neq j$, $u \in K^*$. Let W (resp. $\tilde{W}$) be the subgroup of G (resp. $\tilde{G}$) generated by all $w_{ij}(u)$ (resp. $\tilde{w}_{ij}(u)$); similarly, define H and $\tilde{H}$. (This differs slightly from the notation of Chapter IV.) Note that $\tilde{w}_{ij}(u)^{-1} = \tilde{w}_{ij}(-u)$; but nothing obvious can be said about $\tilde{h}_{ij}(u)^{-1}$.

We already know that H is normal in W, with $W/H \cong S_n$. In fact, there is a semidirect product decomposition of W in $GL(n,K)$: each $w \in W$ is uniquely the product of a permutation matrix corresponding to a permutation π and a diagonal matrix $\mathrm{diag}(u_1,\ldots,u_n)$ of determinant ± 1. Our object now is to study $\tilde{W}$ and its action by conjugation on the generators of $\tilde{G}$.

PROPOSITION. Let $\tilde{w} \in \tilde{W}$, and suppose $\phi(\tilde{w})$ determines the permutation π and the matrix $\mathrm{diag}(u_1,\ldots,u_n)$ as above. Then $\tilde{w}\, \tilde{x}_{ij}(t) \cdot \tilde{w}^{-1} = \tilde{x}_{\pi i,\, \pi j}(u_i u_j^{-1} t)$.

Before proving this, we note an immediate corollary: $A = \mathrm{Ker}(\phi | \tilde{W})$ is central in $\tilde{G}$. It suffices to show that $\tilde{w} \in A$ commutes with all $\tilde{x}_{ij}(t)$, which is clear from the proposition since $\phi(\tilde{w}) = 1$ forces $\pi = 1$, $\mathrm{diag}(u_1,\ldots,u_n) = 1$.

Proof. It is enough to look at a typical generator $\tilde{w} = \tilde{w}_{k\ell}(u)$, for which $\pi = (k\ell)$, $u_k = -u^{-1}$, $u_\ell = u$, $u_m = 1$ otherwise. There are seven cases to consider, depending on how (i,j) and (k,ℓ) are related. We shall indicate how the arguments go, leaving some cases for the reader to complete. To simplify notation, we carry out the calculations in G; but we use only (R1) and (R2), so the results remain valid in $\tilde{G}$.

(1) $i = k$, $j \neq \ell$: Set $v = u^{-1}$. Here $(\pi i, \pi j) = (\ell, j)$ and $u_i u_j^{-1} = -v$. Now $w_{i\ell}(u)\, x_{ij}(t)\, w_{i\ell}(-u) = x_{i\ell}(u)\, x_{\ell i}(-v)\, x_{i\ell}(u)\cdot x_{ij}(t)\, x_{i\ell}(-u)\, x_{\ell i}(v)\, x_{i\ell}(-u) = x_{i\ell}(u)\, x_{\ell i}(-v)\, x_{ij}(t)\, x_{\ell i}(v)\, x_{i\ell}(-u)$ [since $x_{i\ell}(u)$ commutes with $x_{ij}(t)$ and $x_{i\ell}(u)\, x_{i\ell}(-u) = 1$] $= x_{i\ell}(u)\, x_{ij}(t)\, x_{\ell j}(-vt)\, x_{i\ell}(-u)$ [since $x_{\ell i}(-v)\, x_{ij}(t)\, x_{\ell i}(v) = x_{ij}(t)\cdot x_{\ell j}(-vt)$] $= x_{ij}(t)\, x_{i\ell}(u)\, x_{\ell j}(-vt)\, x_{i\ell}(-u) = x_{ij}(t)\, x_{ij}(-uvt)\, x_{\ell j}(-vt) = x_{ij}(t)\, x_{ij}(-t)\, x_{\ell j}(-vt) = x_{\ell j}(-vt)$.

(2) $j = k$, $i \neq \ell$: $w_{j\ell}(u)\, x_{ij}(t)\, w_{j\ell}(-u) = x_{i\ell}(-tu)$.

(3) $j = \ell$, $i \neq k$: $w_{kj}(u)\, x_{ij}(t)\, w_{kj}(-u) = x_{ik}(u^{-1}t)$.

(4) $i = \ell$, $j \neq k$: $w_{ki}(u)\, x_{ij}(t)\, w_{ki}(-u) = x_{kj}(ut)$.

(5) $i = k$, $j = \ell$: $w_{ij}(u)\, x_{ij}(t)\, w_{ij}(-u) = w_{ij}(u)\, (x_{iq}(t), x_{qj}(1))\, w_{ij}(-u)$ [where $n \geq 3$ allows us to select an index $q \neq i,j$] $= (w_{ij}(u)\, x_{iq}(t)\, w_{ij}(-u),\ w_{ij}(u)\, x_{qj}(1)\, w_{ij}(-u)) = (x_{iq}(-tu^{-1}),\ x_{qj}(u^{-1})) = x_{ji}(-tu^{-1})$.

(6) $j = k$, $i = \ell$: $w_{ji}(u)\, x_{ij}(t)\, w_{ji}(-u) = x_{ji}(-tu^2)$.

(7) i,j,k,ℓ all distinct: $w_{k\ell}(u)\, x_{ij}(t)\, w_{k\ell}(-u) = x_{ij}(t)$. □

As a consequence of the proposition and the definitions of $\tilde{w}_{ij}(u)$, $\tilde{h}_{ij}(u)$, we also obtain some formulas governing conjugation in $\tilde{W}$; from (c) we get in particular the fact that $\tilde{H}$ <u>is normal in</u> $\tilde{W}$:

(a) $\tilde{w}\, \tilde{w}_{ij}(u)\, \tilde{w}^{-1} = \tilde{w}_{\pi i,\pi j}(u_i u_j^{-1} u)$ if $\phi(\tilde{w})$ determines π, $\mathrm{diag}(u_1,\ldots,u_n)$ as above.

(b) $\tilde{w}_{ij}(u) = \tilde{w}_{ji}(-u^{-1})$: take $\tilde{w}$ in (a) to be $\tilde{w}_{ij}(u)$.

(c) $\tilde{w}\, \tilde{h}_{ij}(u)\, \tilde{w}^{-1} = \tilde{h}_{\pi i,\pi j}(u_i u_j^{-1} u)\, \tilde{h}_{\pi i,\pi j}(u_i u_j^{-1})$.

(d) $(\tilde{h}_{12}(u), \tilde{h}_{13}(v)) = \tilde{h}_{13}(uv)\, \tilde{h}_{13}(u)^{-1}\, \tilde{h}_{13}(v)^{-1}$: apply (c) with $\tilde{w} = \tilde{h}_{12}(u)$, so $\pi = 1$, $u_1 = u$, $u_2 = u^{-1}$, $\tilde{h}_{12}(u)\, \tilde{h}_{13}(v)\, \tilde{h}_{12}(u)^{-1} = \tilde{h}_{13}(uv)\, \tilde{h}_{13}(u)^{-1}$; then multiply both sides on the right by $\tilde{h}_{13}(v)^{-1}$.

18.4 <u>Steinberg symbols</u>

The calculations just performed lead to some elements of $\tilde{H}$ which are obviously in $\text{Ker } \phi$. To each pair $u, v \in K^*$ is associated a <u>Steinberg symbol</u> $\{u,v\} = (\tilde{h}_{12}(u), \tilde{h}_{13}(v)) = \tilde{h}_{13}(uv)\, \tilde{h}_{13}(u)^{-1}\tilde{h}_{13}(v)^{-1}$. These are elements of the central subgroup $A = \text{Ker}(\phi|\tilde{W})$, which will be shown in 18.5 to generate A. In 18.6 it will be seen that in fact $A = \text{Ker}\phi$.

There is nothing special about the indices used here: for any pair (j,k) of distinct indices, we assert that $\{u,v\} = \tilde{h}_{jk}(uv)\, \tilde{h}_{jk}(u)^{-1}\, \tilde{h}_{jk}(v)^{-1}$. To see this, first suppose $j \neq 1,3$. Then a quick computation yields: $\tilde{w}_{1j}(-1)\, \tilde{h}_{13}(u)^e\, \tilde{w}_{1j}(1) = \tilde{h}_{j3}(u)^e$ $(e = \pm 1)$, or (*) $\tilde{w}_{1j}(-1)\, \tilde{h}_{13}(u)^e = \tilde{h}_{j3}(u)^e\, \tilde{w}_{1j}(-1)$. Since $\{u,v\}$ is central in $\tilde{G}$, we get (using (*)): $\{u,v\} = \tilde{w}_{1j}(-1)\, \{u,v\}\, \tilde{w}_{1j}(1) = \tilde{w}_{1j}(-1)\, \tilde{h}_{j3}(uv) \cdot \tilde{h}_{13}(u)^{-1}\, \tilde{h}_{13}(v)^{-1}\, \tilde{w}_{1j}(1) = \tilde{h}_{j3}(uv)\, \tilde{h}_{j3}(u)^{-1}\, \tilde{h}_{j3}(v)^{-1}\, \tilde{w}_{1j}(-1)\, \tilde{w}_{1j}(1)$ $= \tilde{h}_{j3}(uv)\, \tilde{h}_{j3}(u)^{-1}\, \tilde{h}_{j3}(v)^{-1}$. A similar argument, based on $\tilde{w}_{k3}(1) \cdot \tilde{h}_{j3}(u)^e = \tilde{h}_{jk}(u)^e\, \tilde{w}_{k3}(1)$, shows that $(1,3)$ may be replaced by (j,k). (The cases $j = 1$, $j = 3$ are left to the reader, e.g., replace $(1,3)$ by $(2,3)$ and then replace $(2,3)$ by $(1,2)$.)

To proceed further, we require additional relations in $\tilde{H}$.

<u>PROPOSITION</u>. (a) <u>For distinct indices</u> i,j,k, $\tilde{h}_{jk}(u) = \tilde{h}_{ik}(u) \cdot \tilde{h}_{ij}(u)^{-1}$.

(b) <u>If</u> $j \neq k$, $\tilde{h}_{jk}(u)\, \tilde{h}_{kj}(u) = 1$.

<u>Proof</u>. (a) Using relation (c) at the end of 18.3, we have: $\tilde{w}_{jk}(1)\, \tilde{h}_{ik}(u)^{-1}\, \tilde{w}_{jk}(-1) = \tilde{w}_{jk}(1)\, \tilde{w}_{ik}(1)\, \tilde{w}_{ik}(-u)\, \tilde{w}_{jk}(-1) = \tilde{w}_{ij}(1)\, \tilde{w}_{jk}(1) \cdot \tilde{w}_{ik}(-u)\, \tilde{w}_{jk}(-1) = \tilde{w}_{ij}(1)\, \tilde{w}_{ij}(-u)\, \tilde{w}_{jk}(1)\, \tilde{w}_{jk}(-1) = (\tilde{w}_{ij}(u)\, \tilde{w}_{ij}(-1))^{-1}$ $= \tilde{h}_{ij}(u)^{-1}$, whence $(\tilde{h}_{ik}(u), \tilde{w}_{jk}(1)) = \tilde{h}_{ik}(u)\, \tilde{h}_{ij}(u)^{-1}$. A similar treatment of the first (rather than last) three terms of this commutator yields $(\tilde{h}_{ik}(u), \tilde{w}_{jk}(1)) = \tilde{h}_{jk}(u)$.

(b) Choose $i \neq j,k$, and use part (a) to write $\tilde{h}_{kj}(u) = \tilde{h}_{ij}(u) \cdot \tilde{h}_{ik}(u)^{-1}$, by switching k and j. Multiply this with the equality given by (a) to get $\tilde{h}_{jk}(u)\, \tilde{h}_{kj}(u) = 1$. □

18.5 Determination of A

PROPOSITION. $A = \mathrm{Ker}(\phi|W)$ is included in $\tilde{H}$, and is generated by the Steinberg symbols $\{u,v\}$.

Proof. This is divided into three steps, the first of which will show that $A \subset \tilde{H}$.

(i) Since $\tilde{h}_{ij}(u) = \tilde{w}_{ij}(u)\,\tilde{w}_{ij}(-1)$ by definition, the coset $\tilde{w}_{ij}$ of $\tilde{w}_{ij}(u)$ modulo $\tilde{H}$ is independent of the choice of u. We require several facts about coset multiplication:

(1) $\tilde{w}_{ij}\,\tilde{w}_{1\ell} = \tilde{w}_{\pi 1,\pi\ell}\,\tilde{w}_{ij}$, where $\pi = (ij)$ and $i \neq 1$. (Use (a) at the end of 18.3.)

(2) $\tilde{w}_{1\ell}\,\tilde{w}_{1\ell} = 1$ (since $\tilde{w}_{1\ell}(u) = \tilde{w}_{1\ell}(-u)^{-1}$).

(3) $\tilde{w}_{ij} = \tilde{w}_{ji}$.

(4) $\tilde{w}_{ij}\,\tilde{w}_{1\ell} = \tilde{w}_{1\ell}\,\tilde{w}_{\ell j}$ whenever $\ell \neq j$.

Now let $\tilde{w}$ be some element of A. Its coset mod $\tilde{H}$ is of the form $\prod \tilde{w}_{ij}$, and using (3) we may assume $i < j$ in each term. We wish to move all terms $\tilde{w}_{1\ell}$ to the left. Using (4) we can move $\tilde{w}_{1\ell}$ past $\tilde{w}_{1j}$. Using (1), we can move $\tilde{w}_{1\ell}$ past $\tilde{w}_{ij}$ at the cost of changing $\tilde{w}_{1\ell}$ to $\tilde{w}_{1m}$. So we can rewrite the product so as to have all $\tilde{w}_{1\ell}$ at the left, with ℓ increasing left to right. Then (2) allows us to assume that each $\tilde{w}_{1\ell}$ occurs at most once. Since $\phi(\tilde{w}) = 1$, the associated permutation is 1; but if $\tilde{w}_{1\ell}$ occurs in $\tilde{w}$, the associated permutation (1ℓ) moves 1 to ℓ while the later terms never move ℓ back to 1. So in fact no $\tilde{w}_{1\ell}$ now occurs. Repeating the argument, we eventually get $\tilde{w} \equiv 1 \pmod{\tilde{H}}$.

(ii) With $\tilde{w} \in A$ as above, we may now write $\tilde{w} = \prod \tilde{h}_{ij}(u_{ij})$. Proposition 18.4 allows us to rewrite this as a product of terms $\tilde{h}_{1j}(u_j)$ and inverses of such: $\tilde{h}_{ik}(u) = \tilde{h}_{1k}(u)\,\tilde{h}_{i1}(u)$ for $i,k \neq 1$, while $\tilde{h}_{i1}(u) = \tilde{h}_{1i}(u)^{-1}$.

(iii) Now let A_o be the subgroup of A generated by Steinberg symbols. With $\tilde{w}$ as above, we must show that $\tilde{w} \equiv 1 \pmod{A_o}$. This requires a couple of further relations (cf. the discussion at the beginning of 18.4):

$$(5) \quad \tilde{h}_{1\ell}(uv) \equiv \tilde{h}_{1\ell}(v)\, \tilde{h}_{1\ell}(u) \pmod{A_o},$$

$$(6) \quad \tilde{h}_{1j}(u)\, \tilde{h}_{1\ell}(v) \equiv \tilde{h}_{1\ell}(v)\, \tilde{h}_{1j}(u) \pmod{A_o}.$$

From step (ii) we have $\tilde{w} = \prod \tilde{h}_{1j}(u_j)^{e_j}$ $(e_j = \pm 1)$. Using (5) and (6), the product can be rewritten mod A_o with increasing j and with no repetition of j. But $\phi(\tilde{h}_{1j}(u)^e)$ is a diagonal matrix with u^{-e} in the jth position. So $\phi(\tilde{w}) = 1$ forces all $u_j = 1$, whence $\tilde{w} \equiv 1 \pmod{A_o}$, as desired. □

18.6 Determination of Ker ϕ

THEOREM. Kerϕ = A; *in particular,* Ker ϕ *is central in* G *and is generated by Steinberg symbols.*

Proof. Again we proceed in three steps. The idea is to imitate in $\tilde{G}$ the Bruhat decomposition of G, then to apply Proposition 18.2 in order to force Ker $\phi \subset \tilde{W}$.

(i) First we find a more efficient set of generators for $\tilde{G}$. From (R2) it follows that $\tilde{U}$ is generated by the various $\tilde{x}_{i,i+1}(t)$, e.g., $\tilde{x}_{14}(t) = (\tilde{x}_{12}(t), (\tilde{x}_{23}(1), \tilde{x}_{34}(1))$. So $\tilde{G}$ is generated by the $\tilde{x}_{ij}(t)$, $j = i \pm 1$. Moreover, from Proposition 18.3 we have $\tilde{x}_{i+1,i}(t) = \tilde{w}_{i,i+1}(1)\, \tilde{x}_{i,i+1}(-t)\, \tilde{w}_{i,i+1}(1)^{-1}$. Combining, we see that generators of $\tilde{U}$ together with all $\tilde{w}_{i,i+1}(1)$ suffice to generate G.

(ii) The key step is to show that $\tilde{G} = \tilde{U}\tilde{W}\tilde{U}$. Since the set $\tilde{U}\tilde{W}\tilde{U}$ contains a set of generators for $\tilde{G}$ and is closed under taking inverses, as well as closed under multiplication by $\tilde{U}$, it will be enough to show that it is closed under right multiplication by elements of the form $\tilde{w}_{i,i+1}(1)$.

Start with arbitrary $\tilde{u}_1\, \tilde{w}\, \tilde{u}_2 \in \tilde{U}\tilde{W}\tilde{U}$. Write $\tilde{u}_2 = \tilde{x}_{ij}(t)\, \tilde{u}_3$, where $j = i + 1$ and $\tilde{u}_3$ does not involve any factor $\tilde{x}_{ij}(u)$. Thus $\tilde{u}_3\, \tilde{w}_{ij}(1) = \tilde{w}_{ij}(1)\, \tilde{u}_4$, where $\tilde{u}_4 \in \tilde{U}$. It clearly suffices now to verify that $\tilde{w}\, \tilde{x}_{ij}(t)\, \tilde{w}_{ij}(1) \in \tilde{U}\tilde{W}\tilde{U}$. We may as well assume $t \neq 0$.

Let π be the permutation corresponding to $\phi(\tilde{w})$.

In the good case, $\pi i < \pi j$. Then $\tilde{w}\, \tilde{x}_{ij}(t)\, \tilde{w}_{ij}(1) = \tilde{x}_{\pi i,\pi j}(s) \cdot \tilde{w}\, \tilde{w}_{ij}(1)$ for suitable s, with the first factor lying in $\tilde{U}$.

But suppose instead $\pi i > \pi j$. By definition, since $t \neq 0$, $\tilde{x}_{ij}(t) = \tilde{w}_{ij}(t)\, \tilde{x}_{ij}(-t)\, \tilde{x}_{ji}(t^{-1})$. Thus $\tilde{w}\, \tilde{x}_{ij}(t)\, \tilde{w}_{ij}(1) = \tilde{w}\, \tilde{w}_{ij}(t) \cdot \tilde{x}_{ij}(-t)\, \tilde{x}_{ji}(t^{-1})\, \tilde{w}_{ij}(1) = \tilde{x}_{\pi j,\pi i}(r)\, \tilde{w}\, \tilde{w}_{ij}(t)\, \tilde{w}_{ij}(1)\, \tilde{x}_{ij}(s)$ for suitable r,s. The outer terms both lie in $\tilde{U}$, the inner terms in $\tilde{W}$.

(iii) Thanks to step (ii), a typical element of $\operatorname{Ker}\phi$ can be written as $\tilde{u}_1\, \tilde{w}\, \tilde{u}_2$, with $\tilde{u}_1, \tilde{u}_2 \in \tilde{U}$, $\tilde{w} \in \tilde{W}$. Here $\phi(\tilde{w}) = \phi(\tilde{u}_1)^{-1} \cdot \phi(\tilde{u}_2)^{-1}$. The left side is a monomial matrix, while the right side is upper unitriangular. So both sides equal 1, forcing $\tilde{w} \in A$ (so $\tilde{w}$ is central in $\tilde{G}$), $\tilde{u}_2\, \tilde{u}_1 \in \operatorname{Ker}\phi$. In turn, Proposition 18.2 forces $\tilde{u}_2\, \tilde{u}_1 = 1$. Conclusion: $\tilde{u}_1\, \tilde{w}\, \tilde{u}_2 = \tilde{u}_1\, \tilde{u}_2\, \tilde{w} = \tilde{w} \in A$. □

18.7 Universal property

Let us denote $A = \operatorname{Ker}\phi$ by $\pi_1(G)$ and call it the fundamental group of G. We next show that $\tilde{G}$ is a universal central extension of G, provided K is not too small.

THEOREM. Assume K has at least 5 elements. Given any central extension $1 \to A \to E \to G \to 1$, there is a unique homomorphism $\sigma: \tilde{G} \to E$ making the diagram commute:

$$\begin{array}{ccccccccc} 1 & \to & \pi_1(G) & \to & \tilde{G} & \xrightarrow{\phi} & G & \to & 1 \\ & & \downarrow & & \downarrow \sigma & & \| & & \\ 1 & \to & A & \to & E & \to & G & \to & 1 \end{array}$$

Proof. To define σ, it will be enough to choose elements $\sigma(\tilde{x}_{ij}(t))$ in E which satisfy relations (R1) and (R2). The key to making this choice is the following observation: Given $x, y \in G$ and arbitrary pre-images $x', y' \in E$, the commutator (x',y') in E depends only on x,y (since a central factor from A can be cancelled with its inverse). So begin with arbitrary pre-images $h', x' \in E$ of $h_{ij}(a)$, $x_{ij}(t)$ respectively; (h',x') is then a pre-image of $(h_{ij}(a),$

$x_{ij}(t)) = x_{ij}(a^2t)\ x_{ij}(-t) = x_{ij}((a^2-1)t)$, and is independent of the choices made. Since K is big enough, we can find $a \in K^*$ so that $c = a^2 - 1 \neq 0$. Then ct ranges over K as t does. Relative to this choice of a, we can now define unambiguously: $\sigma(\tilde{x}_{ij}(ct)) = \hat{x}_{ij}(ct) = (h',x')$. We see at the same time that σ will be unique provided it actually defines a homomorphism $\tilde{G} \to E$.

Now define $\hat{w}_{ij}(u)$, $\hat{h}_{ij}(u)$ $(u \in K^*)$ in the obvious way. In particular: $\hat{x}_{ij}(ct) = (\hat{h}_{ij}(a), \hat{x}_{ij}(t))$. We proceed in several steps to verify that the $\hat{x}_{ij}(t)$ satisfy (R1) and (R2).

(1) If $h \in H$, and $\hat{h}$ is an arbitrary pre-image in E, we claim that $\hat{h}\,\hat{x}_{ij}(t)\,\hat{h}^{-1} = \hat{x}_{ij}(dt)$, where $h\,x_{ij}(t)\,h^{-1} = x_{ij}(dt)$, $d \in K^*$. By definition, $\hat{h}\,\hat{x}_{ij}(ct)\,\hat{h}^{-1} = \hat{h}\,(\hat{h}_{ij}(a), \hat{x}_{ij}(t))\,\hat{h}^{-1} = (\hat{h}\,\hat{h}_{ij}(a)\,\hat{h}^{-1}, \hat{h}\,\hat{x}_{ij}(t)\,\hat{h}^{-1}) = (\hat{h}_{ij}(a), \hat{x}_{ij}(dt)) = \hat{x}_{ij}(cdt)$, using the fact that the commutator is independent of the choice of pre-images. Since ct ranges over K as t does, our claim is proved.

(2) If $(x_{ij}(t), x_{k\ell}(u)) = 1$ in G for all $t, u \in K$, we assert that $(\hat{x}_{ij}(t), \hat{x}_{k\ell}(u)) = 1$ in E. (This is a special case of (R2).)

Set $f(t,u) = (\hat{x}_{ij}(t), \hat{x}_{k\ell}(u)) \in A$. We have to show that $f(t,u) = 1$ for all $t, u \in K$. Observe that f is "additive" in each variable: using the commutator identity $(xy,z) = (x,(y,z))\ (y,z)\ (x,z)$, we calculate $f(t+t',u) = (\hat{x}_{ij}(t+t'), \hat{x}_{k\ell}(u)) = (\hat{x}_{ij}(t)\hat{x}_{ij}(t'),\hat{x}_{k\ell}(u)) = (\hat{x}_{ij}(t), (\hat{x}_{ij}(t'), \hat{x}_{k\ell}(u))\quad (\hat{x}_{ij}(t'), \hat{x}_{k\ell}(u))\ (\hat{x}_{ij}(t), \hat{x}_{k\ell}(u)) = (\hat{x}_{ij}(t'), \hat{x}_{k\ell}(u))\ (\hat{x}_{ij}(t), \hat{x}_{k\ell}(u)) = f(t',u)\ f(t,u) = f(t,u)\ f(t',u)$. Here we used the fact that $(x_{ij}(t'), x_{k\ell}(u))$ is central in E.

Now we have to consider the various possible cases in which it happens that $(x_{ij}(t), x_{k\ell}(u)) = 1$.

(a) $(i,j,k,\ell$ all distinct) Here $h_{ij}(v)$ commutes with all $x_{k\ell}(u)$, so (1) above forces the same to happen for the chosen pre-images in E. For $v \in K^*$, consider the effect of conjugating the (central) element $f(t,u)$ by $\hat{h}_{ij}(v)$: $f(t,u) = (\hat{h}_{ij}(v)\ \hat{x}_{ij}(t)\hat{h}_{ij}(v)^{-1}, \hat{x}_{k\ell}(u)) = (\hat{x}_{ij}(v^2t),\hat{x}_{k\ell}(u)) = f(v^2t,u)$. Using additivity, we deduce $1 = f(v^2t,u)\ f(t,u)^{-1} = f(v^2t,u)\ f(-t,u) = f((v^2-1)t,u)$. If we choose v so that $v^2-1 \neq 0$, this forces $f(t,u) = 1$ for all t,u.

(b) $(i = k,\ j \neq \ell$, or else $i \neq k,\ j = \ell)$ Here $h_{ij}(v)\ x_{k\ell}(u)\cdot h_{ij}(v)^{-1} = x_{k\ell}(vu)$. Conjugate $f(t,u)$ by $\hat{h}_{ij}(v^2)\ \hat{h}_{k\ell}(v^{-1})$ to obtain $f(t,u) = f(tv^3,u)$. Find $v \in K^*$ for which $v^3-1 \neq 0$, then conclude as in case (a) that $1 = f(t,u)$.

(c) $(i = k,\ j = \ell)$ Using the fact that $n \geq 3$, find an index $q \neq i,j$ and conjugate by $\hat{h}_{iq}(v)$ to get (*) $f(t,u) = f(vt,vu)$. Using the fact that K has at least 5 elements, find $v \in K^*$ for which $v \neq v^2$, $1 - v + v^2 \neq 0$. Then (*) forces $f(t(v - v^2),u) = f(t, u/(v - v^2)) = f(t, (u/v) + (u/(1 - v))) = f(t,u/v)\ f(t,u/(1 - v))$ [by additivity] $= f(vt,u)\ f((1 - v)t,u)$ [by (*)] $= f(t,u)$ [by additivity]. It follows as before that $f(t,u) = 1$.

(3) We can now verify (R1) for the $\hat{x}_{ij}(t)$: Set $\hat{x} = \hat{x}_{ij}(t/c)\cdot \hat{x}_{ij}(u/c)\ \hat{x}_{ij}((t + u)/c)^{-1} \in A$. Conjugate by $\hat{h}_{ij}(a)$, using $(\hat{h}_{ij}(a), \hat{x}_{ij}(t)) = \hat{x}_{ij}(ct)$ and step (2), to get $\hat{x} = \hat{x}\ \hat{x}_{ij}(t)\ \hat{x}_{ij}(u)\ \hat{x}_{ij}(t+u)^{-1}$. Then divide both sides by $\hat{x}$.

(4) Step (2) already verifies some cases of (R2). It remains to check that when $i \neq \ell$, $(\hat{x}_{ij}(t), \hat{x}_{j\ell}(u)) = \hat{x}_{i\ell}(tu)$. Clearly the left side equals $f(t,u)\ \hat{x}_{i\ell}(tu)$ for some $f(t,u) \in A$. As in step (2), it has to be shown that all $f(t,u) = 1$.

First check that this f is also additive in each variable:

$f(t+t',u)\ \hat{x}_{i\ell}((t+t')u) = (\hat{x}_{ij}(t+t'), \hat{x}_{j\ell}(u)) = (\hat{x}_{ij}(t)\ \hat{x}_{ij}(t'),\hat{x}_{j\ell}(u))$
$= (\hat{x}_{ij}(t),(\hat{x}_{ij}(t'),\hat{x}_{j\ell}(u)))\ (\hat{x}_{ij}(t'),\hat{x}_{j\ell}(u))\ (\hat{x}_{ij}(t),\hat{x}_{j\ell}(u))$
$= (\hat{x}_{ij}(t),\ f(t',u)\ \hat{x}_{i\ell}(t,u))\ f(t',u)\ \hat{x}_{i\ell}(t'u)\ f(t,u)\ \hat{x}_{i\ell}(tu)$
$= f(t,u)\ f(t',u)\ \hat{x}_{i\ell}((t+t')u)$, using (R1) and the fact that $(\hat{x}_{ij}(t), \hat{x}_{i\ell}(t'u)) = 1$ by step (2), while $f(t',u)$ is central. Similarly, f is additive in the other variable.

Now imitate the argument in part (a) of step (2): conjugate by

$\hat{h}_{ij}(v^2)\,\hat{h}_{j\ell}(v)$ to get $f(t,u) = f(v^3t,u)$. Choose $v \in K^*$ such that $v^3 - 1 \neq 0$, and as before obtain $f(t,u) = 1$ for all t,u. □

Remark. There really are exceptions to the theorem when K is too small, cf. Steinberg [1].

18.8 Properties of Steinberg symbols

We saw in 18.6 that $\pi_1(G)$ is generated by Steinberg symbols $\{u,v\} = (\tilde{h}_{12}(u), \tilde{h}_{13}(v)) = \tilde{h}_{13}(uv)\,\tilde{h}_{13}(u)^{-1}\,\tilde{h}_{13}(v)^{-1}$, where $u,v \in K^*$. (Other pairs of subscripts could equally well be used.) These symbols have some striking properties, notably:

(1) $\{u,v\} = \{v,u\}^{-1}$ (skew-symmetry), so that $\{u,u\}^2 = 1$.

(2) $\{u_1u_2,v\} = \{u_1,v\}\{u_2,v\}$ and $\{u,v_1v_2\} = \{u,v_1\}\{u,v_2\}$ (bilinearity).

(3) $\{u,1-u\} = 1$ provided the symbol is defined (i.e., provided $u,\ 1-u \in K^*$).

The proofs of these properties go as follows:

For (1), note that $\{v,u\}^{-1} = (\tilde{h}_{12}(v), \tilde{h}_{13}(u))^{-1} = (\tilde{h}_{13}(u),\tilde{h}_{12}(v))$ $= \{u,v\}$ due to our freedom in choosing subscripts.

In view of (1), the first part of (2) will imply the second. We require the commutator identity: (*) $(a_1a_2,b) = (a_1,(a_2,b))(a_2,b)(a_1,b)$. Now $\{u_1u_2,v\} = (\tilde{h}_{12}(u_1u_2), \tilde{h}_{13}(v))$

$= (\tilde{h}_{12}(u_1)\,\tilde{h}_{12}(u_2)\,\{u_1,u_2\}, \tilde{h}_{13}(v))$

$= (\tilde{h}_{12}(u_1)\,\tilde{h}_{12}(u_2), \tilde{h}_{13}(v))$ [since $\{u_1,u_2\}$ is central]

$= (\tilde{h}_{12}(u_1),(\tilde{h}_{12}(u_2),\tilde{h}_{13}(v))(\tilde{h}_{12}(u_2),\tilde{h}_{13}(v))(\tilde{h}_{12}(u_1),\tilde{h}_{13}(v))$ [by (*)]

$= 1\,\{u_2,v\}\{u_1,v\}$ [since $\{u_2,v\}$ is central]

$= \{u_1,v\}\{u_2,v\}$.

To prove (3), write $v = 1-u$. We have to show that $1 = \{u,v\}$ $= \tilde{h}_{13}(uv)\,\tilde{h}_{13}(u)^{-1}\,\tilde{h}_{13}(v)^{-1} = \tilde{w}_{13}(uv)\tilde{w}_{13}(-1)\tilde{w}_{13}(1)\tilde{w}_{13}(-u)\tilde{w}_{13}(1)\tilde{w}_{13}(-v)$ or: $\tilde{w}_{13}(uv) = \tilde{w}_{13}(v)\,\tilde{w}_{13}(-1)\,\tilde{w}_{13}(u)$. In the right hand side, substitute $\tilde{w}_{31}(1) = \tilde{w}_{13}(-1) = \tilde{x}_{31}(1)\,\tilde{x}_{13}(-1)\,\tilde{x}_{31}(1)$. Then use

$\tilde{w}_{13}(v)\ \tilde{x}_{31}(1) = \tilde{x}_{13}(-v^2)\ \tilde{w}_{13}(v)$ and $\tilde{x}_{31}(1)\ \tilde{w}_{13}(u) = \tilde{w}_{13}(u)\ \tilde{x}_{13}(-u^2)$, to obtain $\tilde{x}_{13}(-v^2)\ \tilde{w}_{13}(v)\ \tilde{x}_{13}(-1)\ \tilde{w}_{13}(u)\ \tilde{x}_{13}(-u^2)$. This equals (by definition)

$\tilde{x}_{13}(-v^2)\ \tilde{x}_{13}(v)\ \tilde{x}_{31}(-v^{-1})\ \tilde{x}_{13}(v)\ \tilde{x}_{13}(-1)\ \tilde{x}_{13}(u)\ \tilde{x}_{31}(-u^{-1})\tilde{x}_{13}(u)\tilde{x}_{13}(-u^2)$ $= \tilde{x}_{13}(v-v^2)\ \tilde{x}_{31}(-v^{-1}-u^{-1})\tilde{x}_{13}(u-u^2) = \tilde{x}_{13}(uv)\ \tilde{x}_{31}(-1/uv)\ \tilde{x}_{13}(uv) =$ $\tilde{w}_{13}(uv)$. □

More generally, let us define a <u>Steinberg symbol</u> with values in an abelian group A to be a bilinear function $c:K^* \times K^* \to A$ satisfying $c(u,1-u) = 1$ whenever $u,1-u \in K^*$. So the function $K^* \times K^* \to \pi_1(G)$ sending (u,v) to $\{u,v\}$ is an example. We do not need to assume explicitly that c is skew-symmetric, as the following result shows.

<u>PROPOSITION. Let</u> $c:K^* \times K^* \to A$ <u>be an A-valued Steinberg symbol. Then</u> $c(u,v) = c(v,u)^{-1}$ <u>and</u> $c(u,-u) = 1$. <u>Moreover, if</u> K <u>is finite, then</u> $c(u,v) = 1$ <u>for all</u> u,v.

<u>Proof</u>. Note that the bilinearity forces $c(1,u) = 1$ for all u. Next we show that $c(u,-u) = 1$. We may assume $u \neq 1$. From bilinearity we get $c(u,v)^{-1} = c(u^{-1},v)$. Since $(1-u)/(1-u^{-1}) = -u$ for $u \neq 1$, we then find $1 = c(u,1-u)\ c(u^{-1},1-u^{-1}) = c(u,1-u)\ c(u,1-u^{-1})^{-1}$ $= c(u,1-u)\ c(u,(1-u^{-1})^{-1}) = c(u,(1-u)/(1-u^{-1})) = c(u,-u)$.

To prove skew-symmetry, we use $1 = c(u,-u)$ to write $c(u,v) =$ $c(u,v)\ c(u,-u) = c(u,-uv) = c(u,-uv)\ c(uv,-uv)^{-1} = c(u,-uv)\ c(1/uv,-uv)$ $= c(v^{-1},-uv) = c(v^{-1},-uv)\ c(v^{-1},-v^{-1}) = c(v^{-1},u) = c(v,u^{-1})$.

Now let K be <u>finite</u>. Then K^* is cyclic, say generated by v. From skew-symmetry we know that $c(v,v)^2 = 1$. On the other hand, from bilinearity we know that $c(v^i,v^j) = c(v,v)^{ij}$. So it will suffice to show that $c(v,v) = 1$, or just that an <u>odd</u> power of $c(v,v) = 1$. When char $K = 2$, $c(v,v) = c(v,-v) = 1$, so we are done. Suppose char K is odd, so $|K^*| = q-1$ is even and exactly half the elements of K^* are squares. Consider the map $u \mapsto 1-u$ from $K^* - \{1\}$ onto itself. If all $(q-1)/2$ nonsquares map to (distinct!) squares, we would have too many squares, since 1 is also a square. So there exists a nonsquare u_0 for which $1-u_0$ is also a nonsquare: say $u_0 = v^i$, $1-u_0 = v^j$ for odd i,j. Then $1 = c(u_0,1-u_0) = c(v,v)^{ij}$ as desired. □

Note that, in conjunction with Theorem 18.6, the last statement implies a simple presentation of the finite group $SL(n,K)$, $n \geq 3$, using the relations (R1) and (R2).

Exercise. (a) If K is only assumed to be an algebraic extension of a finite field, prove that all Steinberg symbols $c: K^* \times K^* \to A$ are trivial.

(b) If c is a Steinberg symbol, $c(u,u^{-1})^2 = 1$.

Remark. Symplectic groups yield Steinberg symbols as above, but these can fail to be bilinear when a "long root" is involved. So Matsumoto has to give a more general definition of what he calls "Steinberg cocycles". Nonsymplectic groups all behave like SL_n $(n \geq 3)$, yielding only bilinear symbols.

§19. Matsumoto's theorem

In the preceding section we saw that the study of central extensions of $SL(n,K)$ $(n \geq 3)$ leads to Steinberg symbols. Matsumoto [1] showed, conversely, that for a given Steinberg symbol $c: K^* \times K^* \to A$, there exists a corresponding central extension. His method applies equally well to other Chevalley groups (with some extra complications in the symplectic case); for a briefer treatment emphasizing just $SL(n,K)$, cf. Milnor [1, §11-12]. We will present Matsumoto's arguments here. They involve a mixture of both depth and persistence, and require a certain amount of involvement on the part of the reader.

19.1 Central extensions and cocycles

First we want to recall some standard homological machinery. In this subsection only, G can be an arbitrary group. Consider a central extension $1 \to A \to E \to G \to 1$. Choose arbitrary pre-images $\bar{x} \in E$ for the elements $x \in G$, subject only to the requirement that $\bar{1} = 1$. Then given $x, y \in G$, $\bar{x}\,\bar{y} = f(x,y)\,\overline{xy}$ for some $f(x,y) \in A$. The resulting function $f : G \times G \to A$ is called the <u>cocycle</u> of the extension (relative to the chosen liftings). From the normalization $\bar{1} = 1$ and the associative law in E we quickly deduce:

(1) $f(1,y) = 1 = f(x,1)$,

(2) $f(x,y)\, f(xy,z) = f(y,z)\, f(x,yz)$.

Conversely, if $f:G \times G \to A$ is a function satisfying (1), (2), it arises as a cocycle for a central extension of G with kernel A, constructed as follows. Set $E = A \times G$, with group operation: $(a,x)\cdot(a',x') = (aa'\, f(x,x'),\, xx')$. It is routine to check that E is a group, with central subgroup $\{(a,1)\} \cong A$ and resulting quotient isomorphic to G. The lifting $x \mapsto (1,x)$ yields the given cocycle f.

If we vary the choice of liftings, the cocycle varies only by a <u>coboundary</u>, a function $g:G \times G \to A$ satisfying $g(x,y) = h(x)\,h(y)\cdot h(xy)^{-1}$ for some function $h:G \to A$ with $h(1) = 1$. Cocycles may be multiplied by multiplying their values, and then form an abelian group, having the set of coboundaries as a subgroup, with resulting quotient group denoted $H^2(G,A)$. Elements of this group correspond 1-1 with equivalence classes of central extensions of G by A, if we define extensions (with groups E,E') to be <u>equivalent</u> when there exists a homomorphism $E \to E'$ (necessarily an isomorphism) making the diagram commute:

$$\begin{array}{ccccccccc} 1 & \to & A & \to & E & \to & G & \to & 1 \\ & & \| & & \downarrow & & \| & & \\ 1 & \to & A & \to & E' & \to & G & \to & 1 \end{array}$$

In 18.8 we defined an A-valued <u>Steinberg symbol</u> $c:K^* \times K^* \to A$ to be a bilinear function satisfying $c(u,1-u) = 1$. The set $S(K^*,A)$ of all Steinberg symbols forms an abelian group if we multiply symbols by multiplying their values in A. Bilinearity implies that a Steinberg symbol is a cocycle in the above sense, so it defines a central extension of K^* by A. In 19.3 we shall use a given c to build a central extension of the diagonal subgroup of $SL(n,K)$, which is a product of copies of K^*.

19.2 Statement of the theorem

Again let $G = SL(n,K)$, $n \geq 3$. Say $|K| > 4$. Given a central extension $1 \to A \to E \to G \to 1$, Steinberg's theorem 18.7 provides a unique homomorphism σ making the diagram commute:

$$\begin{array}{ccccccccc} 1 & \to & \pi_1(G) & \to & \tilde{G} & \to & G & \to & 1 \\ & & \downarrow\sigma & & \downarrow\sigma & & \| & & \\ 1 & \to & A & \to & E & \to & G & \to & 1 \end{array}$$

Here $\pi_1(G)$ is generated by symbols $\{u,v\}$ (see 18.6). The restriction of σ to $\pi_1(G)$ yields a Steinberg symbol $c \in S(K^*,A)$ defined by $c(u,v) = \sigma\{u,v\}$. In this way we obtain natural maps:

$$H^2(G,A) \to \mathrm{Hom}(\pi_1(G),A) \to S(K^*,A).$$

Observe that these homomorphisms are injective: If $\sigma\{u,v\} = 1$ for all u,v, $\sigma(\tilde{G}) \cong G$ splits the given central extension. The second map is injective because the symbols $\{u,v\}$ generate $\pi_1(G)$.

THEOREM (Matsumoto). *Given $c \in S(K^*,A)$, there exists a central extension of $G = SL(n,K)$, $n \geq 3$, with kernel A and with resulting Steinberg symbol c. Thus $H^2(G,A) \cong \mathrm{Hom}(\pi_1(G),A) \cong S(K^*,A)$.*

The proof will occupy 19.3 - 19.6. Note that K may be assumed to be infinite, since for finite K, $S(K^*,A)$ is trivial (18.8). To lighten the notation somewhat, we shall just write down the proof when $n = 3$. (This case is already adequate to settle the congruence subgroup problem affirmatively when $K = \mathbb{Q}$, as remarked at the end of 17.5.) The argument in the general case is entirely similar. We denote by α, $-\alpha$, β, $-\beta$ the respective pairs $(1,2)$, $(2,1)$, $(2,3)$, $(3,2)$.

To construct a central extension of G, we shall define a central extension of the diagonal subgroup H in 19.3, then extend the construction to the monomial group W in 19.5 (via an important auxiliary construction in 19.4) and finally conclude in 19.6.

Exercise. Deduce from Matsumoto's theorem that $\pi_1(G)$ is generated by the symbols $\{u,v\}$ $(u,v \in K^*)$ subject only to the relations:

$$\{u_1u_2,v\}\ \{u_1,v\}^{-1}\ \{u_2,v\}^{-1} = 1,$$

$$\{u,v_1v_2\}\ \{u,v_1\}^{-1}\ \{u,v_2\}^{-1} = 1,$$

$$\{u,1-u\} = 1.$$

[Take A to be the abstract group so generated, let c be the resulting A-valued Steinberg symbol, and compare the resulting central extension with $1 \to \pi_1(G) \to \tilde{G} \to G \to 1$.]

19.3 The diagonal group

We are given $c: K^* \times K^* \to A$, and wish to define a cocycle $f: H \times H \to A$. Each $h \in H$ has a unique decomposition $h = h_\alpha(u)h_\beta(v)$. So we just define $f(h_\alpha(u_\alpha)\ h_\beta(u_\beta),\ h_\alpha(v_\alpha)\ h_\beta(v_\beta)) = c(u_\alpha,v_\alpha)c(u_\beta,v_\beta)\cdot c(v_\alpha,u_\beta)$. It is routine to check that f satisfies conditions (1), (2) in 19.1, hence gives rise to a central extension $1 \to A \to \hat{H} \to H \to 1$, in the manner described there. The prescribed lift of $h_\gamma(u)$ $(\gamma = \alpha,\beta)$ will be denoted $\hat{h}_\gamma(u)$. We have the following explicit rules of multiplication and commutation in $\hat{H}$:

$$\hat{h}_\gamma(u)\ \hat{h}_\gamma(v) = c(u,v)\ \hat{h}_\gamma(uv)$$

$$(\hat{h}_\gamma(u),\ \hat{h}_\gamma(v)) = c(u,v^2) = c(u^2,v)$$

$$(\hat{h}_\alpha(u),\ \hat{h}_\beta(v)) = c(u,v^{-1}) = c(u,v)^{-1} = c(u^{-1},v) = c(v,u)$$

$$(\hat{h}_\alpha(u),\ \hat{h}_\alpha(v)\ \hat{h}_\beta(v)) = c(u,v).$$

As an abstract group, $\hat{H}$ is generated by all $a \in A$, $\hat{h}_\alpha(u)$, $\hat{h}_\beta(v)$, subject only to the relations on A along with $(a,-) = 1$ and $\hat{h}_\gamma(u)\ \hat{h}_\gamma(v) = c(u,v)\ \hat{h}_\gamma(uv)$ $(\gamma = \alpha,\beta)$, $(\hat{h}_\alpha(u),\ \hat{h}_\beta(v)) = c(u,v^{-1})$.

19.4 An auxiliary construction

In order to construct a suitable central extension $1 \to A \to \hat{W} \to$

$W \to 1$, whose restriction to H is $\hat{H}$, we need to study closely the behavior of a crucial finite subgroup of W. Set $w_\alpha = w_\alpha(-1) = \begin{pmatrix} 0 & -1 & 0 \\ 1 & 0 & 0 \\ 0 & 0 & 1 \end{pmatrix}$, $w_\beta = w_\beta(-1) = \begin{pmatrix} 1 & 0 & 0 \\ 0 & 0 & -1 \\ 0 & 1 & 0 \end{pmatrix}$, $h_\alpha = w_\alpha^2$, $h_\beta = w_\beta^2$. Let $\mathcal{W}$ be the subgroup of W generated by w_α, w_β, $\mathcal{H}$ the subgroup of H generated by h_α, h_β. Direct examination makes it plain that $\mathcal{H}$ is a Klein 4-group, with $\mathcal{H} = \mathcal{W} \cap H$ normal in $\mathcal{W}$ and $\mathcal{W}/\mathcal{H} \cong S_3$. (Exercise. $\mathcal{W}, \mathcal{H}$ are the respective intersections of W, H with $SL(3,\mathbb{Z})$.)

PROPOSITION. All relations involving the generators w_α, w_β of $\mathcal{W}$ are consequences of the following ones:

(W1) $h_\alpha w_\beta h_\alpha^{-1} = w_\beta^{-1}$, $h_\beta w_\alpha h_\beta^{-1} = w_\alpha^{-1}$

(W2) $w_\alpha w_\beta w_\alpha^{-1} = w_\beta w_\alpha^{-1} w_\beta^{-1}$, $w_\beta w_\alpha w_\beta^{-1} = w_\alpha w_\beta^{-1} w_\alpha^{-1}$

(W3) $h_\alpha^2 = 1 = h_\beta^2$.

Proof. First note that these relations do hold in $\mathcal{W}$. So there is an epimorphism $\phi\colon \overline{\mathcal{W}} \to \mathcal{W}$ from the group $\overline{\mathcal{W}}$ generated by elements $\overline{w}_\alpha$, $\overline{w}_\beta$ subject only to (W1) - (W3), where $\overline{h}_\alpha = \overline{w}_\alpha^{\,2}$ and $\overline{h}_\beta = \overline{w}_\beta^{\,2}$. From (W1) we deduce that the subgroup $\overline{\mathcal{H}}$ of $\overline{\mathcal{W}}$ generated by $\overline{h}_\alpha$, $\overline{h}_\beta$ is normal. For example, $\overline{w}_\beta^{-1}\, \overline{h}_\alpha \overline{w}_\beta = \overline{w}_\beta^{-2}\, \overline{h}_\alpha = \overline{h}_\beta^{-1}\, \overline{h}_\alpha$. Combining (W1) with (W3), we deduce that $\overline{\mathcal{H}}$ is abelian: $\overline{h}_\alpha\, \overline{h}_\beta\, \overline{h}_\alpha^{-1}\, \overline{h}_\beta^{-1} = \overline{h}_\alpha \quad \overline{w}_\beta^{2}\, \overline{h}_\alpha^{-1}\, \overline{h}_\beta^{-1} = (\overline{h}_\alpha\, \overline{w}_\beta\, \overline{h}_\alpha^{-1})(\overline{h}_\alpha\, \overline{w}_\beta\, \overline{h}_\alpha^{-1})\overline{h}_\beta^{-1} = \overline{w}_\beta^{-1}\, \overline{w}_\beta^{-1}\, \overline{h}_\beta^{-1} = \overline{h}_\beta^{-1}\, \overline{h}_\beta^{-1} = 1$. This implies at once that $|\overline{\mathcal{H}}| = |\mathcal{H}| = 4$. Now $\overline{\mathcal{W}}/\overline{\mathcal{H}}$ is generated by a pair of involutions a,b, which (thanks to (W2)) satisfy $aba = bab$. This is easily seen to be a presentation of S_3, so $\overline{\mathcal{W}}/\overline{\mathcal{H}}$ has order at most 6. This shows that $|\overline{\mathcal{W}}| = |\mathcal{W}|$, so ϕ is 1-1 . □

Remark. In place of (W1), (W2), we could just as well use the relations:

(W1') $w_\beta^{-1} h_\alpha w_\beta = h_\alpha h_\beta$, $w_\alpha^{-1} h_\beta w_\alpha = h_\beta h_\alpha$,

(W2') $w_\alpha w_\beta w_\alpha = w_\beta w_\alpha w_\beta$.

We leave the verification as an exercise. Now we can formulate the main result of this section.

THEOREM. *Let $\hat{W}$ be the group generated by elements $\hat{w}_\alpha$, $\hat{w}_\beta$ subject only to the relations* (W1), (W2) *above (or equivalently,* (W1'), (W2')). *Let $\hat{H}$ be the subgroup generated by $\hat{h}_\alpha = \hat{w}_\alpha^2$, $\hat{h}_\beta = \hat{w}_\beta^2$. Denote by π the canonical epimorphism $\hat{W} \to W$. Then:*

(a) $Z = \operatorname{Ker} \pi$ *lies in the center of $\hat{W}$ and is generated by* $\hat{h}_\alpha^2$, $\hat{h}_\beta^2$.

(b) Z *has order* 2.

(c) $\hat{H}$ *has defining relations:* $\hat{h}_\beta^{-1} \hat{h}_\alpha \hat{h}_\beta = \hat{h}_\alpha \hat{h}_\beta^2$ *and* $\hat{h}_\alpha^{-1} \cdot \hat{h}_\beta \hat{h}_\alpha = \hat{h}_\beta \hat{h}_\alpha^2$.

Proof. Note (as in the proof of the preceding proposition) that (W1) forces $\hat{H}$ to be normal in $\hat{W}$. It will be necessary to construct an explicit model of $\hat{W}$, a group of order 48, containing the subgroup $\hat{H}$ of order 8. (The latter turns out to be the familiar quaternion group.) We proceed in four steps.

(1) First we observe that the elements $\hat{h}_\alpha^2$, $\hat{h}_\beta^2$ of Z are central in $\hat{W}$, i.e., commute with the two generators: $\hat{h}_\alpha^2 \hat{w}_\beta \hat{h}_\alpha^{-2} = \hat{h}_\alpha (h_\alpha \hat{w}_\beta \hat{h}_\alpha^{-1}) \hat{h}_\alpha^{-1} = \hat{h}_\alpha \hat{w}_\beta^{-1} \hat{h}_\alpha^{-1} = \hat{w}_\beta$, by two applications of (W1). Obviously, $\hat{h}_\alpha^2$ commutes with $\hat{w}_\alpha$. Now the quotient group $W/\langle \hat{h}_\alpha^2, \hat{h}_\beta^2 \rangle$ satisfies (W1)-(W3) and maps onto W, hence is isomorphic to W thanks to the proposition above. It follows that Z is generated by $\hat{h}_\alpha^2$, $\hat{h}_\beta^2$, and also that $\hat{W}/\hat{H} \cong S_3$.

(2) Next we check that $\hat{H}$ actually satisfies the relations given in (c). For this we use (W1') twice: $\hat{h}_\beta^{-1} \hat{h}_\alpha \hat{h}_\beta = \hat{w}_\beta^{-1} (\hat{w}_\beta^{-1} \hat{h}_\alpha \hat{w}_\beta) \cdot \hat{w}_\beta = \hat{w}_\beta^{-1} (\hat{h}_\alpha \hat{h}_\beta) \hat{w}_\beta = \hat{h}_\alpha \hat{h}_\beta \hat{w}_\beta^{-1} \hat{h}_\beta \hat{w}_\beta = \hat{h}_\alpha \hat{h}_\beta^2$ (and similarly for the other relation).

(3) Let T be the abstract group generated by elements a, b

subject only to the relations in (c): $b^{-1}ab = ab^2$, $a^{-1}ba = ba^2$. We deduce: $b^2 = a^{-1}b^{-1}ab = (b^{-1}a^{-1}ba)^{-1} = a^{-2}$. So b^2 commutes with a, and a^2 commutes with b. In turn: $a = b^{-2}ab^2 = b^{-1}(b^{-1}ab)b$ $= b^{-1}(ab^2)b = (b^{-1}ab)b^2 = (ab^2)b^2 = ab^4$, forcing $b^4 = 1$. Similarly, $a^4 = 1$. Therefore, $a^2 = b^2$. (So Z is in fact cyclic, generated by $\hat{h}_\alpha{}^2 = \hat{h}_\beta{}^2$.)

On the other hand, it is well known that the quaternion group of order 8 has a presentation $\langle A,B \mid A^4 = 1 = B^4,\ A^2 = B^2,\ B^{-1}AB = A^{-1}\rangle$, from which the relations (c) are immediate. So we conclude that T is the quaternion group. In view of steps (1), (2), we have $|\hat{H}| \leq 8$, $|Z| \leq 2$. All assertions of the theorem will follow, once we prove that Z has order at least 2.

(4) Now we must construct an actual group (of order 48) isomorphic to $\hat{W}$, having a normal subgroup isomorphic to T. In particular, we have to specify how the generators of $\hat{W}$ are to act on T. Define specific automorphisms of T as follows:

$$\theta_a: a \mapsto a,\quad b \mapsto ba^{-1}, \qquad \theta_a{}^{-1}: a \mapsto a,\quad b \mapsto ba,$$

$$\theta_b: a \mapsto ab^{-1},\quad b \mapsto b, \qquad \theta_b{}^{-1}: a \mapsto ab,\quad b \mapsto b\ .$$

It is an easy exercise, using the relations in T, to show that θ_a, θ_b really are automorphisms, e.g., check that the relations hold when the pair a,b is replaced by the pair a, ba^{-1}. Note that $\theta_a{}^2$ sends b to aba^{-1}, so agrees with conjugation by a in T (similarly for $\theta_b{}^2$). Now consider the subgroup of Aut T generated by θ_a, θ_b . We claim that relations (W1)-(W3) all hold here. For example, (W1) requires $\theta_a{}^2\,\theta_b\theta_a{}^{-2} = \theta_b{}^{-1}$, which is true since each side does the same thing to a,b. We leave the routine verifications to the reader.

Having made these preliminary observations about Aut T, we are ready to introduce our candidate for $\hat{W}$. This will be a group X of permutations of the set $T \times S_3$. Take $s_a = (12)$ and $s_b = (23)$ as distinguished generators of S_3, so each element has a corresponding "length" $\ell(s)$ (cf. §12) between 0 and 3. Define a map $\lambda_a : T \times S_3 \to T \times S_3$ by the recipe:

$$\lambda_a(t,s) = \begin{cases} (\theta_a(t), & s_a s) \quad \text{if} \quad \ell(s_a s) > \ell(s) \\ (\theta_a(t)a, & s_a s) \quad \text{otherwise.} \end{cases}$$

Similarly, define λ_b. We must check that λ_a, λ_b are actually permutations of $T \times S_3$ (i.e., are 1-1). Suppose $\lambda_a(t,s) = \lambda_a(t',s')$. Clearly $s_a s = s_a s'$, so $s = s'$. In turn, $\theta_a(t) = \theta_a(t')$, forcing $t = t'$ since θ_a is 1-1. Now take X to be the group of permutations generated by λ_a, λ_b. We want to verify the relations (W1), (W2), or rather the equivalent pair (W1), (W2').

For (W1) it must be checked that, e.g., $\lambda_a^2 \lambda_b \lambda_a^{-2} = \lambda_b^{-1}$, or $\lambda_b \lambda_a^2 \lambda_b = \lambda_a^2$. Use the fact that $\lambda_a^2(t,s) = (at,s)$, which follows from $\theta_a^2(t) = ata^{-1}$. Then consider the case when $\ell(s_b s) < \ell(s)$:

$\lambda_b\lambda_a^2 \lambda_b(t,s) = \lambda_b\lambda_a^2(\theta_b(t)b,\ s_b s) = \lambda_b(a\theta_b(t)b,\ s_b s)$

$= (\theta_b(a)\theta_b^2(t)b,\ s) \qquad [\text{since } \ell(s_b s_b s) = \ell(s) > \ell(s_b s)]$

$= (ab^{-1}btb^{-1}b,\ s) \qquad [\text{since } \theta_b^2(t) = btb^{-1},\ \ \theta_b(a) = ab^{-1}]$

$= (at,\ s) = \lambda_a^2(t,s)$. The case when $\ell(s_b s) > \ell(s)$ is treated similarly.

For (W2') we need to check: $\lambda_a \lambda_b \lambda_a = \lambda_b\lambda_a \lambda_b$. Apply both sides to a typical (t,s) and treat the 6 possible choices of s (rather than dealing in a more sophisticated way with the length function, as would be required if S_n or another such Weyl group were involved). For example:

$$(t,s_a s_b) \xrightarrow[\lambda_a]{} (\theta_a(t)a,\ s_b) \xrightarrow[\lambda_b]{} (\theta_b\theta_a(t)ab^{-1}b,1) \xrightarrow[\lambda_a]{} (\theta_a\theta_b\theta_a(t)a,s_a),$$

$$(t,s_a s_b) \xrightarrow[\lambda_b]{} (\theta_b(t),s_b s_a s_b) \xrightarrow[\lambda_a]{} (\theta_a\theta_b(t)a,s_b s_a) \xrightarrow[\lambda_b]{} (\theta_b\theta_a\theta_b(t)ab^{-1}b,s_a).$$

Then use the fact that $\theta_a \theta_b \theta_a = \theta_b \theta_a \theta_b$. Similarly:

$$(t,s_b) \xrightarrow[\lambda_b]{} (\theta_a(t),s_a s_b) \xrightarrow[\lambda_a]{} (\theta_b\theta_a(t)ba^{-1},s_a) \xrightarrow[\lambda_b]{} (\theta_b\theta_a\theta_b(t)bab,s_b s_a).$$

But $bab = a$ in T, so $\lambda_a\lambda_b\lambda_a$ agrees here with $\lambda_b\lambda_a\lambda_b$.

We noted already that $\lambda_a^2(t,s) = (at,s)$ and $\lambda_b^2(t,s) = (bt,s)$. It follows at once that the subgroup of X generated by λ_a^2, λ_b^2 is isomorphic to T. By (W1) or (W1'), this subgroup is normal. Since X satisfies (W1) and (W2), it follows that X is a homomorphic image of $\hat{\mathcal{W}}$, with $\hat{H}$ mapping onto the copy of T in X. In particular, $\hat{H}$ has order at least 8, which is all we needed to know. (In turn, $\hat{H} \cong T$ and $\hat{\mathcal{W}} \cong X$. So we have an explicit model of $\hat{\mathcal{W}}$.) □

Remark. The ideas in this proof carry over almost unchanged to other Chevalley groups of rank ≥ 2, with one important modification: for symplectic groups, Z is infinite cyclic. This reflects the fact (noted by Matsumoto [1, p.36]) that $\hat{\mathcal{W}}$ is realizable in all cases as the inverse image of $\mathcal{W}$ in the universal covering group of the corresponding Lie group over $\mathbb{R}$. It is well known that $Sp(2n,\mathbb{R})$ has infinite cyclic fundamental group, while $SL(n,\mathbb{R})$, $n \geq 3$, and other groups have fundamental group of order 2. Milnor [1] uses an approach like this for $SL(n,K)$, to avoid explicitly constructing a model of $\hat{\mathcal{W}}$.

19.5 The monomial group

Now we combine the central extensions $\hat{H} \xrightarrow{\phi} H$ and $\hat{\mathcal{W}} \xrightarrow{\pi} \mathcal{W}$ constructed in 19.3, 19.4, to obtain an extension $\hat{W} \to W$ compatible with ϕ. The idea is to begin with a semidirect product $\hat{\mathcal{W}} \times \hat{H}$ ($\hat{H}$ normal), then define $\hat{W}$ to be a suitable quotient. More precisely, we shall construct a commutative square:

$$\begin{array}{ccc} \hat{\mathcal{W}} \times \hat{H} & \xrightarrow{\quad q \quad} & \hat{W} \\ {\scriptstyle (\pi,\phi)} \downarrow & & \downarrow {\scriptstyle \phi} \\ \mathcal{W} \times H & \xrightarrow[\quad p \quad]{} & W \end{array}$$

Here $\phi: \hat{W} \to W$ will extend $\phi: \hat{H} \to H$, with $\operatorname{Ker}\phi$ still equal to A.

PROPOSITION. There exist automorphisms $\delta_\alpha, \delta_\beta$ of $\hat{H}$ such that: (a) $\delta_\alpha, \delta_\beta$ fix A pointwise, while δ_α^{-1} sends $\hat{h}_\alpha(u)$ to

$\hat{h}_\alpha(u^{-1})$, $\hat{h}_\beta(u)$ to $\hat{h}_\beta(u)\,\hat{h}_\alpha(u)$, δ_β^{-1} sends $\hat{h}_\alpha(u)$ to $\hat{h}_\alpha(u)\,\hat{h}_\beta(u)$, $\hat{h}_\beta(u)$ to $\hat{h}_\beta(u^{-1})$.

(b) By passage to the quotient, δ_α, δ_β induce on H the respective inner automorphisms $\mathrm{Int}\; w_\alpha$, $\mathrm{Int}\; w_\beta$.

(c) $\delta_\alpha^{\,2} = \delta_\alpha^{-2}$ is $\mathrm{Int}\;\hat{h}_\alpha(-1)$, and similarly for δ_β.

(d) δ_α, δ_β satisfy the relations (W1), (W2) of 19.4 defining $\hat{W}$.

Proof. (a) As noted in 19.3, $\hat{H}$ is generated by the elements $a \in A$, $\hat{h}_\alpha(u)$, $\hat{h}_\beta(v)$, subject only to the relations listed there. So we have to check that the images of these generators under $\delta_\alpha, \delta_\beta$ (or their inverses) satisfy the same relations. There is no problem about the relations on A or the relations expressing the fact that A is central. Consider a relation: $\hat{h}_\gamma(u)\,\hat{h}_\gamma(v) = c(u,v)\,\hat{h}_\gamma(uv)$. Taking $\delta = \delta_\alpha^{-1}$, we compute:

$$\begin{aligned}
\delta(\hat{h}_\beta(u))\;\delta(\hat{h}_\beta(v)) &= \hat{h}_\beta(u)\;\hat{h}_\alpha(u)\;\hat{h}_\beta(v)\;\hat{h}_\alpha(v)\\
&= c(u,v^{-1})\;\hat{h}_\beta(u)\;\hat{h}_\beta(v)\;\hat{h}_\alpha(u)\;\hat{h}_\alpha(v)\\
&= c(u,v)^2\;c(u,v^{-1})\;\hat{h}_\beta(uv)\;\hat{h}_\alpha(uv)\\
&= c(u,v)\;\delta(\hat{h}_\beta(uv))\\
&= \delta(c(u,v))\;\delta(\hat{h}_\beta(uv))
\end{aligned}$$

$$\begin{aligned}
\delta(\hat{h}_\alpha(u))\;\delta(\hat{h}_\alpha(v)) &= \hat{h}_\alpha(u^{-1})\;\hat{h}_\alpha(v^{-1})\\
&= c(u^{-1},v^{-1})\;\hat{h}_\alpha((uv)^{-1})\\
&= c(u,v)\;\delta(\hat{h}_\alpha(uv))\\
&= \delta(c(u,v))\;\delta(\hat{h}_\alpha(uv))
\end{aligned}$$

Commutation relations are treated similarly:

$$\begin{aligned}
\delta(\hat{h}_\alpha(u))\;\delta(\hat{h}_\beta(v)) &= \hat{h}_\alpha(u^{-1})\;\hat{h}_\beta(v)\;\hat{h}_\alpha(v)\\
&= c(u^{-1},v^{-1})\;\hat{h}_\beta(v)\;\hat{h}_\alpha(u^{-1})\;\hat{h}_\alpha(v)\\
&= c(u^{-1},v)\;c(u^{-1},v^{-1})\;\hat{h}_\beta(v)\;\hat{h}_\alpha(u^{-1}v)\\
&= \hat{h}_\beta(v)\;\hat{h}_\alpha(u^{-1}v)
\end{aligned}$$

$$\delta(c(u,v^{-1}))\,\delta(\hat{h}_\beta(v))\,\delta(\hat{h}_\alpha(u)) = c(u,v^{-1})\;\hat{h}_\beta(v)\;\hat{h}_\alpha(v)\;\hat{h}_\alpha(u^{-1})$$

$$= c(u,v^{-1})\ c(v,u^{-1})\,\hat{h}_\beta(v)\ \hat{h}_\alpha(vu^{-1})$$

$$= c(u,v^{-1})\ c(u,v)\ \hat{h}_\beta(v)\ \hat{h}_\alpha(vu^{-1})$$

$$= \hat{h}_\beta(v)\ \hat{h}_\alpha(vu^{-1})$$

To conclude that δ_α^{-1}, δ_β^{-1} (hence δ_α, δ_β) define automorphisms of $\hat{H}$, it remains to see that they are 1-1. But this will follow from (c).

(b) By direct computation (cf. the earlier calculations 18.3 in the Steinberg group), $w_\alpha(1)\ h_\beta(u)\ w_\alpha(-1) = h_\beta(u)\ h_\alpha(u)$, $w_\alpha(1)\ h_\alpha(u)\cdot w_\alpha(-1) = h_\alpha(u^{-1})$, etc.

(c) $\delta_\alpha^{-2}(\hat{h}_\alpha(u)) = \hat{h}_\alpha(u) = \text{Int}\ \hat{h}_\alpha(-1)\ (\hat{h}_\alpha(u))$. Also, $\delta_\alpha^{-2}(\hat{h}_\beta(u)) = \hat{h}_\beta(u)\ \hat{h}_\alpha(u)\ \hat{h}_\alpha(u^{-1}) = c(u,u^{-1})\ \hat{h}_\beta(u)$. On the other hand, $\text{Int}\ \hat{h}_\alpha(-1)\ (\hat{h}_\beta(u)) = c(-1,u^{-1})\ \hat{h}_\beta(u)$. So it remains to check that $c(u,u^{-1}) = c(-1,u^{-1})$, using properties of Steinberg symbols: $c(-1,u^{-1})\ c(u,u^{-1})^{-1} = c(-1,u^{-1})\ c(u^{-1},u^{-1}) = c(-u^{-1},u^{-1}) = 1$.

(d) For (W1) we need: $\delta_\alpha^{2}\ \delta_\beta\ \delta_\alpha^{-2} = \delta_\beta^{-1}$, or equivalently, $\delta_\beta^{-1}\ \delta_\alpha^{-2}\ \delta_\beta^{-1} = \delta_\alpha^{-2}$. Apply the left side to $\hat{h}_\alpha(u)$ and use part (c) above to get: $\delta_\beta^{-1}\ \delta_\alpha^{-2}\ \hat{h}_\alpha(u)\ \hat{h}_\beta(u) = \delta_\beta^{-1}(\hat{h}_\alpha(u)\ c(u,u^{-1})\hat{h}_\beta(u)) = \hat{h}_\alpha(u)\ \hat{h}_\beta(u)\ c(u,u^{-1})\ \hat{h}_\beta(u^{-1}) = c(u,u^{-1})^2\ \hat{h}_\alpha(u)\ \hat{h}_\beta(1) = \hat{h}_\alpha(u) = \delta_\alpha^{-2}(\hat{h}_\alpha(u))$. We use here the easily checked fact that $c(u,u^{-1})^2 = 1$.

For (W2) it suffices to check (W2'), or $\delta_\alpha^{-1}\ \delta_\beta^{-1}\ \delta_\alpha^{-1} = \delta_\beta^{-1}\cdot \delta_\alpha^{-1}\ \delta_\beta^{-1}$. Apply the left side to $\hat{h}_\alpha(u)$ to get: $\delta_\alpha^{-1}\ \delta_\beta^{-1}\ (\hat{h}_\alpha(u^{-1})) = \delta_\alpha^{-1}(\hat{h}_\alpha(u^{-1})\ \hat{h}_\beta(u^{-1})) = \hat{h}_\alpha(u)\ \hat{h}_\beta(u^{-1})\ \hat{h}_\alpha(u^{-1}) = c(u^{-1},u^{-1})\ \hat{h}_\alpha(u)\cdot \hat{h}_\alpha(u^{-1})\ \hat{h}_\beta(u^{-1}) = c(u^{-1},u^{-1})\ c(u,u^{-1})\ \hat{h}_\beta(u^{-1}) = \hat{h}_\beta(u^{-1})$. Applying the right side to $\hat{h}_\alpha(u)$ yields: $\delta_\beta^{-1}\ \delta_\alpha^{-1}\ (\hat{h}_\alpha(u).\hat{h}_\beta(u)) = \delta_\beta^{-1}\ (\hat{h}_\alpha(u^{-1})\cdot \hat{h}_\beta(u)\ \hat{h}_\alpha(u)) = c(u,u)\ \delta_\beta^{-1}\ (\hat{h}_\alpha(u^{-1})\ \hat{h}_\alpha(u)\ \hat{h}_\beta(u)) = c(u,u)\ c(u^{-1},u)\cdot \delta_\beta^{-1}(\hat{h}_\beta(u)) = \hat{h}_\beta(u^{-1})$. □

From (a) and (d) we obtain an action of $\hat{W}$ on $\hat{H}$, allowing us to form the semidirect product $\hat{W} \times \hat{H}$. This maps by (π,ϕ) onto the semidirect product $W \times H$ (where W acts on H by inner automomorphisms in W). From (a) and (b) we see that $\phi : \hat{H} \to H$ is equivariant relative to the respective actions of $\hat{W}, W$; so (π,ϕ) is a group homomorphism.

There is an obvious epimorphism $p : W \times H \to W$, having kernel of order 4 consisting of all pairs $(h,h^{-1}) \in W \times H$ with $h \in H$. We imitate this by defining $J \subset \hat{W} \times \hat{H}$ as the set of all pairs $(\hat{h},j(\hat{h})^{-1})$, where $j : \hat{H} \to \hat{H}$ is defined by $j(\hat{h}_\alpha) = \hat{h}_\alpha(-1)$, $j(\hat{h}_\beta) = \hat{h}_\beta(-1)$. (Existence of such a group homomorphism j follows from the presentation of $\hat{H}$ contained in part (c) of Theorem 19.4: $\hat{h}_\alpha(-1)$, $\hat{h}_\beta(-1)$ satisfy the same relations in $\hat{H}$.) It is readily checked that J is a normal subgroup, allowing us to define $\hat{\mathbb{W}} = (\hat{W} \times \hat{H})/J$, with canonical map $q:\hat{W} \times \hat{H} \to \hat{\mathbb{W}}$. It is now clear that we have the desired commutative square, with $\phi: \hat{\mathbb{W}} \to \mathbb{W}$ extending $\phi: \hat{H} \to H$ and $\operatorname{Ker}\phi = A$.

19.6 Conclusion of the proof

Now we must fit the pieces together, to obtain the required central extension $1 \to A \to E \to G \to 1$ having Steinberg symbol c. We know from 18.7 that E must have a subgroup mapped isomorphically onto U, and that E must then look something like $U \hat{\mathbb{W}} U$ ($\hat{\mathbb{W}}$ as in 19.5). But it is not feasible to construct E directly in this spirit. We proceed indirectly, first constructing a set S in 1-1 correspondence with the desired group, then defining E to be a group of permutations of S, acting simply transitively on S.

To describe S we need precise information about the $\mathbb{W}$-component in the Bruhat decomposition $G = U \mathbb{W} U$ (cf. §12, where the letter N was used in place of $\mathbb{W}$). When we write $g = uwu'$, the element w is uniquely determined; call it $\nu(g)$. Besides the "simple roots" α, β, we need the other "positive root" $\gamma = (1,3)$ below.

LEMMA. Let $g \in G$. Then: (a) $\nu(w_\alpha(-1)\, g) = w_\alpha(-1)\, \nu(g)$ or $h_\alpha(u^{-1})\, \nu(g)$ for some u; (b) $\nu(g\, w_\beta(1)) = \nu(g)\, w_\beta(1)$ or $\nu(g) h_\beta(v)$ for some v; and similarly for α, β interchanged.

Proof. It will suffice to prove (a) in detail. We rely on the fact that a typical element of U can be written as (say) $x_\alpha(s)\cdot x_\beta(t)\ x_\gamma(u)$, and that conjugation by $w_\alpha(-1)$ keeps the latter two factors in U. Thus it is enough to consider the case $g = x_\alpha(u)\ w$ ($w \in W$), $u \neq 0$. With w is associated a permutation π. Two cases arise.

(i) $\pi(\alpha)$ is a positive root. This means that $w_\alpha(-1)\ g = w_\alpha(-1)\cdot x_\alpha(u)\ w = w_\alpha(-1)\ w\ x$, where $x \in U$. Accordingly, $\nu(w_\alpha(-1)\ g) = w_\alpha(-1)\ w = w_\alpha(-1)\ \nu(g)$.

(ii) $\pi(\alpha)$ is not positive. Here we invoke the definition: $x_\alpha(u) = x_{-\alpha}(u^{-1})\ x_\alpha(-u)\ w_\alpha(u)$. Thus $w_\alpha(-1)\ g = w_\alpha(-1)\ x_\alpha(u)\ w = w_\alpha(-1)\cdot x_{-\alpha}(u^{-1})\ x_\alpha(-u)\ w_\alpha(u)\ w$. Now conjugation by $w_\alpha(u)\ w$ takes $x_\alpha(-u)$ back into U, while conjugation by $w_\alpha(-1)$ takes $x_{-\alpha}(u^{-1})$ into U. After rewriting we obtain: $x\ w_\alpha(-1)\ w_\alpha(u)\ w\ x'$ ($x,x' \in U$), with $w_\alpha(-1)\ w_\alpha(u) = h_\alpha(u^{-1})$. So $\nu(w_\alpha(-1)\ g) = h_\alpha(u^{-1})\ \nu(g)$. □

With $\nu:G \to W$ as above and $\phi:\hat{W} \to W$ as in 19.5, let S be the set of all pairs $(g,\hat{w}) \in G \times \hat{W}$ satisfying: $\nu(g) = \phi(\hat{w})$. We shall construct a group E of permutations of S and show that it acts simply transitively. Generators for E are given as follows:

(1) If $\hat{h} \in \hat{H}$, define $\lambda(\hat{h})\ (g,\hat{w}) = (\phi(\hat{h})\ g,\ \hat{h}\ \hat{w})$.

(2) If $x \in U$, define $\lambda(x)\ (g,\hat{w}) = (xg,\ \hat{w})$.

(3) Define $\lambda_\alpha(g,\hat{w})$ to be either $(w_\alpha(-1)g,\hat{w}_\alpha(-1)\ \hat{w})$ or else $(w_\alpha(-1)\ g,\ \hat{h}_\alpha(u^{-1})\ \hat{w})$ depending on how $\nu(w_\alpha(-1)\ g)$ appears in the lemma above. Define λ_β similarly.

In each case it has to be observed that the maps take S into itself. For (1), (2), this follows from: $\nu(xg) = \nu(g)$, $\nu(h\ g) = h\ \nu(g)$ if $g \in G$, $x \in U$, $h \in H$. For (3) we appeal to the lemma.

It also has to be observed that the maps are permutations of S. This is clear for (1), (2), where the inverses are given explicitly by $\lambda(\hat{h}^{-1})$, $\lambda(x^{-1})$. It is less obvious for (3). But notice that $\lambda_\alpha^{\ 2} = \lambda(\hat{h}_\alpha(-1))$, and similarly for $\lambda_\beta^{\ 2}$. So λ_α, λ_β are also permutations.

It is obvious that the group homomorphisms $\lambda: \hat{H} \to E$, $\lambda: U \to E$ are monomorphisms. Moreover, by restricting permutations of S to the first coordinate we obtain a natural homomorphism ψ from E to the group of permutations of G induced by the various left multiplications by elements of H and U, together with $w_\alpha(-1)$, $w_\beta(-1)$. These elements generate G, and G is isomorphic to the resulting permutation group (Cayley), so ψ may be viewed as an epimorphism $E \to G$. It is evident that $A \subset \hat{H}$ lies in $\mathrm{Ker}\,\psi$, and that A is central in E. We have to show that $\mathrm{Ker}\,\psi$ is precisely A.

Observe next that E acts <u>transitively</u> on S: If $g, g' \in G$, use the surjectivity of ψ to find $\lambda \in E$ with $\psi(\lambda) = g'g^{-1}$. Then if $(g,\hat{w}) \in S$, $\lambda(g,\hat{w}) = (g'g^{-1}g,\hat{w}') = (g',\hat{w}')$ for some $\hat{w}'$. In turn, all possible choices of second coordinate for the fixed first coordinate g' are obtained by applying the various $\lambda(a)$, $a \in A$, which leave the first coordinate fixed.

<u>KEY LEMMA</u>. E <u>acts simply transitively on</u> S.

Assuming this lemma, we can easily finish the argument: If $\psi(\lambda) = 1$, $\lambda(g,\hat{w}) = (g,\hat{w}') = \lambda(a)\,(g,\hat{w})$ for some $a \in A$, as in the preceding remarks. Simple transitivity forces $\lambda = \lambda(a)$, so $\mathrm{Ker}\psi = A$. Simple transitivity also implies at once that λ_α, λ_β satisfy the relations (W1), (W2) of 19.4, since $\hat{w}_\alpha(-1)$, $\hat{w}_\beta(-1)$ do so in $\hat{W}$. It follows that $\lambda:\hat{H} \to E$ is extendible to a monomorphism $\lambda:\hat{W} \to E$. In turn, it follows from the construction of $\phi:\hat{W} \to W$ that the Steinberg symbol of the resulting central extension $1 \to A \to E \to G \to 1$ is precisely the symbol c with which we began.

<u>Proof of lemma</u>. Here Matsumoto introduces a very clever trick. He defines a second group E^* acting transitively on S, but involving right rather than left action. E^* is generated by permutations defined as follows:

(1) If $\hat{h} \in \hat{H}$, $(g,\hat{w})\ \rho(\hat{h}) = (g\phi(\hat{h}), \hat{w}\ \hat{h})$.

(2) If $x \in U$, $(g,\hat{w})\ \rho(x) = (g\ x,\hat{w})$.

(3) Define ρ_α, ρ_β acting on the right by analogy with the action of λ_α, λ_β (following part (b) of the previous lemma).

Suppose we can show that the two actions commute, in the sense that $[\lambda\ (g,\hat{w})]\ \rho = \lambda[(g,\hat{w})\ \rho]$ for all $\lambda \in E$, $\rho \in E^*$. It will follow that E acts simply transitively on S: $\lambda(g,\hat{w}) = \lambda'(g,\hat{w})$ implies that $\lambda[(g,\hat{w})\rho] = \lambda'[(g,\hat{w})\rho]$ for all $\rho \in E^*$, whence $\lambda = \lambda'$ since E^* acts transitively.

Commutativity of all $\lambda(\hat{h})$, $\lambda(x)$ $(\hat{h} \in \hat{H}, x \in U)$ with all $\rho(\hat{h})$, $\rho(x)$ is immediate from the definitions. So (as usual) the crucial step involves the Weyl group. It is enough to show: $[\lambda_\alpha(g,\hat{w})]\rho_\beta = \lambda_\alpha[(g,\hat{w})\rho_\beta]$. By looking at the Bruhat decomposition of g (cf. proof of previous lemma), we see that it suffices to consider just elements g of the form $x_\alpha(u)\ w(\pi)\ x_\beta(v)$, where $w(\pi)$ is a representative of $\pi \in S_3$ in W, written as a product of factors $w_\alpha(1)$, $w_\beta(1)$. After modifying the second coordinate $\hat{w}$ by a central element $\lambda(a) = \rho(a)$ $(a \in A)$, we can then take $\hat{w}$ to be the corresponding product of factors $\hat{w}_\alpha(1)$, $\hat{w}_\beta(1)$. Then there are several distinct possibilities to analyze:

(i) $\pi(\beta) \neq \pm\ \alpha$. Here λ_α, ρ_β behave "independently" and the result follows easily.

(ii) $\pi(\beta) = \alpha$, so $\pi = (23)(12)$. Here $g = x_\alpha(u)\ w(\pi)\ x_\beta(v) = w(\pi)\ x_\beta(u)\ x_\beta(v)$, so it is enough to treat the case $g = x_\alpha(u)\ w(\pi)$. We have to compare the effects of $\lambda_\alpha, \rho_\beta$ on the second coordinate of $(g,\hat{w})$: $\lambda_\alpha[(g,\hat{w})\rho_\beta]$ has second coordinate (*) $\hat{h}_\alpha(u^{-1})\ \hat{w}(\pi)\ \hat{w}_\beta(-1)$, while $[\lambda_\alpha(g,\hat{w})]\rho_\beta$ yields (**) $\hat{w}_\alpha(-1)\ \hat{w}(\pi)\ \hat{h}_\beta(u)$. Now just compute:

$$
\begin{aligned}
(*) &= \hat{h}_\alpha(u^{-1})\ \hat{w}(\pi)\ \hat{h}_\beta(-1)\ \hat{w}_\beta(1) \\
&= \hat{h}_\alpha(u^{-1})\ \hat{h}_\alpha(-1)\ \hat{w}(\pi)\ \hat{w}_\beta(1) \\
&= c(u^{-1},-1)\ \hat{h}_\alpha(-u^{-1})\ \hat{w}(\pi(23))
\end{aligned}
$$

$$= c(u^{-1},-1)\ \hat{h}_\alpha(-u^{-1})\ \hat{w}_\beta(1)\ \hat{w}_\alpha(1)\ \hat{w}_\beta(1)$$

$$(**)\quad = \hat{h}_\alpha(-1)\ \hat{w}_\alpha(1)\ \hat{h}_\alpha(u)\ \hat{w}(\pi)$$

$$= \hat{h}_\alpha(-1)\ \hat{h}_\alpha(u^{-1})\ \hat{w}_\alpha((12)\pi)$$

$$= c(-1,u^{-1})\ \hat{h}_\alpha(-u^{-1})\ \hat{w}_\alpha(1)\ \hat{w}_\beta(1)\ \hat{w}_\alpha(1).$$

But $c(-1,u^{-1}) = c(-1,u)^{-1} = c(u,-1) = c(u,(-1)^{-1}) = c(u,-1)^{-1} = c(u^{-1},-1)$. Moreover, relation (W2') holds in $\hat{W}$, whence (*) = (**).

(iii) $\pi(\beta) = -\alpha$, so $\pi = (12)(23)(12) = (23)(12)(23)$.

In case u or v is 0, or else $uv = 1$, the argument is similar to the one above. Otherwise, as in (ii), we compare second coordinates:

$$(*)\quad = \hat{h}_\alpha((u-v^{-1})^{-1})\ \ \hat{w}(\pi)\ \hat{h}_\beta(v)$$

$$= \hat{h}_\alpha((u-v^{-1})^{-1})\ \ \hat{h}_\alpha(v^{-1})\ \hat{w}(\pi)$$

$$(**)\quad = \hat{h}_\alpha(u^{-1})\ \hat{w}(\pi)\ \ \hat{h}_\beta(v-u^{-1})$$

$$= \hat{h}_\alpha(u^{-1})\ \hat{h}_\alpha((v-u^{-1})^{-1}\ \hat{w}(\pi).$$

This leads to a computation with Steinberg symbols: $c((u-v^{-1})^{-1}, v^{-1}) \overset{?}{=} c(u^{-1},(v-u^{-1})^{-1})$. Here the left side equals $c(u-v^{-1},v) = c(-(1-uv)v^{-1},v) = c(1-uv,v)\ c(-v,v)^{-1} = c(1-uv,v)$. The right side equals $c(u,-(1-uv)u^{-1}) = c(u,1-uv)\ c(u,-u)^{-1} = c(1-uv,u)^{-1}$. But $c(1-uv,u)\ c(1-uv,v) = c(1-uv,uv) = 1$, so we are done .□

19.7 The big cell

In order to adapt Matsumoto's theorem to the case when K is a topological field, we have to recall the discussion in 12.4 (1), which adapts at once from GL_n to SL_n. In the topological case, the argument there implies that the product map $U \times H \times U^- \to G$ is a homeomorphism onto an open subset Ω of G, called the "big cell". We are writing the factors in reverse order here, to be consistent with Matsumoto [1].

The following lemma, needed in 19.8, shows how to rewrite certain products as elements of the corresponding big cell in the Steinberg group $\tilde{G}$. We state it just for the pair of roots $\pm\alpha$, but it works just as well for any other pair.

LEMMA. Let $\tilde{G} = St(3,K)$ as in 18.1, with $\alpha = (1,2)$, $-\alpha = (2,1)$. If $u, v \in K^*$ and $t = 1+uv \neq 0$, then $\tilde{x}_{-\alpha}(u)\ \tilde{x}_{\alpha}(v) = \tilde{x}_{\alpha}(vt^{-1})\tilde{h}_{\alpha}(t^{-1})\cdot \{u,t\}\ \tilde{x}_{-\alpha}(ut^{-1})$.

Proof. (For notational ease, we carry out the calculations in G, using the relations for $\tilde{G}$.) From the definition of w_α we get $x_{-\alpha}(u) = x_\alpha(u^{-1})\ w_\alpha(-u^{-1})\ x_\alpha(u^{-1})$ since $u \neq 0$. So $x_{-\alpha}(u)\ x_\alpha(v)$ $= x_\alpha(u^{-1})\ w_\alpha(-u^{-1})\ x_\alpha(u^{-1}+v)$; here $u^{-1}+v = tu^{-1}$. Use the same trick to replace $x_\alpha(tu^{-1})$ by $w_\alpha(-tu^{-1})^{-1}\ x_\alpha(-tu^{-1})\ x_{-\alpha}(ut^{-1})$. Insert $w_\alpha(-1)\ w_\alpha(1)$ between the w_α factors and use the definitions of h_α and the Steinberg symbol, along with Proposition 18.3, to obtain:

$$\begin{aligned}
x_{-\alpha}(u)\ x_\alpha(v) &= x_\alpha(u^{-1})\ h_\alpha(-u^{-1})\ h_\alpha(-tu^{-1})^{-1}\ x_\alpha(-tu^{-1})\ x_{-\alpha}(ut^{-1}) \\
&= x_\alpha(u^{-1})\ \{-tu^{-1},t^{-1}\}\ h_\alpha(t^{-1})\ x_\alpha(-tu^{-1})\ x_{-\alpha}(ut^{-1}) \\
&= x_\alpha(u^{-1})\ x_\alpha(-t^{-1}u^{-1})\ h_\alpha(t^{-1})\ \{-tu^{-1},t^{-1}\}\ x_{-\alpha}(ut^{-1}) \\
&= x_\alpha(vt^{-1})\ h_\alpha(t^{-1})\ \{u,t\}\ x_{-\alpha}(ut^{-1}).
\end{aligned}$$

Here we use the fact that $\{-t, t\} = 1$ to compute $\{-tu^{-1},t^{-1}\}=\{-t,t\}^{-1}\cdot \{u^{-1},t\}^{-1} = \{u,t\}$. □

19.8 The topological case

Now let K be an infinite topological field (e.g., a local field), and let A be an abelian (Hausdorff) topological group. As before, $G = SL(3,K)$. An exact sequence $1 \to A \xrightarrow{i} E \xrightarrow{\pi} G \to 1$ is called a topological central extension if E is a (Hausdorff) topological group, i maps A isomorphically onto a closed subgroup of E, and π is a continuous homomorphism inducing an isomorphism of topological groups $E/A \cong G$. (This says that π is a quotient map, in particular an open map.) There is an obvious notion of equivalence, so we obtain $H^2_{top}(G,A)$ and a natural (forgetful) map into $H^2(G,A)$. We will be interested in characterizing the Steinberg symbols which correspond to topological central extensions.

Given a topological central extension as above, with Steinberg symbol c, we make the convention that $c(0,1) = c(1,0) = 1$; therefore, the function $(u,v) \longmapsto c(u,1+uv)$ is defined in a neighborhood of $(0,0)$ in $K \times K$. As in 18.7 we define pre-images of generators such as $x_\alpha(t)$ satisfying: $\hat{x}_\alpha((a^2-1)t) = (\hat{h}_\alpha(a), \hat{x}_\alpha(t))$, where $a^2-1 \neq 0$. Define subgroups of E in the obvious way: $\hat{U}, \hat{U}^-, \hat{W}, \hat{H}, \hat{X}_\alpha$, ...

LEMMA. With notation as above, we have:

(a) For each root γ, $\hat{x}_\gamma : K \to \hat{X}_\gamma$ is an isomorphism of K (as topological group) onto a closed subgroup of E.

(b) $\hat{U}, \hat{U}^-$ are closed in E and map isomorphically onto U, U^-.

(c) For each γ, $\hat{h}_\gamma : K^* \to E$ is a homeomorphism onto a closed subset of E.

(d) The product map $\hat{U} \times \hat{H} \times \hat{U}^- \to E$ is a homeomorphism onto an open subset Ω_E of E.

(e) c is continuous, and the function $K \times K \to A$ given by $(u,v) \longmapsto c(u,1+uv)$ is continuous at $(0,0)$.

Proof. (a) If V is a neighborhood of 1 in E, the continuity of the map $\hat{x} \longmapsto (\hat{h}_\gamma(t), \hat{x})$ yields an open neighborhood V' of 1 in E such that for $x \in V'$, $(\hat{h}_\gamma(t), \hat{x}) \in V$. Since π is open, $\pi(V')$ is an open neighborhood of 1 in G and hence contains all $x_\gamma(u)$ for u ranging over some neighborhood V_0 of 0 in K. Thus for $u \in V_0$, $\hat{x}_\gamma((a^2-1)u) \in V$. It follows that $\hat{x}_\gamma$ is continuous at 0, hence everywhere. In turn it follows easily that $\hat{X}_\gamma$ is closed in E. From (a) we then deduce (b) - (d).

(e) We need Lemma 19.7, which carries over to E via the universal property of the Steinberg group. The map $K \times K \to E$ defined by $(u,v) \longmapsto \hat{x}_{-\alpha}(u)\, \hat{x}_\alpha(v)$ is continuous, thanks to (a), and provided $t = 1+uv \neq 0$, the image lies in $\hat{U}\,\hat{H}\,\hat{U}^-$. Indeed, for $u \neq 0$, 19.7 allows us to rewrite: $\hat{x}_{-\alpha}(u)\, \hat{x}_\alpha(v) = \hat{x}_\alpha(vt^{-1})\, \hat{h}_\alpha(t^{-1})\, c(u,t)\cdot$

$\hat{x}_{-\alpha}(ut^{-1})$. With our convention on c, this equality holds whenever $t \neq 0$. Continuity of $\hat{x}_\alpha$, $\hat{h}_\alpha$, c implies that $c(u,1+uv) \to 1$ as $(u,v) \to (0,0)$. □

The lemma prompts us to call a Steinberg symbol $c: K^* \times K^* \to A$ topological if it is continuous and if the function $(u,v) \mapsto c(u,1+uv)$ is continuous at $(0,0)$, where $c(0,1) = c(1,0) = 1$ by convention. The set of all these symbols, a subgroup of $S(K^*,A)$, is denoted $S_{top}(K^*,A)$. Now we have a square:

$$\begin{array}{ccc} H^2(G,A) & \longleftrightarrow & S(K^*,A) \\ \uparrow & & \uparrow \\ H^2_{top}(G,A) & \longrightarrow & S_{top}(K^*,A) \end{array}$$

Note that the bottom arrow is 1-1: If a topological central extension has trivial Steinberg symbol, the lemma implies that the various $\hat{X}_\gamma$ generate a closed subgroup of E mapped isomorphically onto G. (As a result, the left vertical arrow is also 1-1.)

It remains to be seen that the lower arrow can be reversed.

THEOREM. Let K be an infinite topological field, A an abelian topological group. If $c \in S_{top}(K^*,A)$, the corresponding central extension given by Theorem 19.2 has the structure of a topological central extension in such a way that c is the resulting Steinberg symbol.

Proof. First we must topologize E suitably. Start with the analogue of the big cell: $\hat{U}$, $\hat{U}^-$ (as above) can be endowed with the respective topologies of U, U^-, while $\hat{H}$ can be topologized by requiring that the map $H \times A \to \hat{H}$ sending $(h_\alpha(s)\,h_\beta(t), a)$ to $\hat{h}_\alpha(s)\,\hat{h}_\beta(t)\,a$ be a homeomorphism. Then the product map takes $\hat{U} \times \hat{H} \times \hat{U}^-$ bijectively onto $\Omega_E = \hat{U}\,\hat{H}\,\hat{U}^-$, which is given the product topology. Now decree a neighborhood of 1 in E to be a subset of E which intersects Ω_E in a neighborhood of 1. It has to be checked that

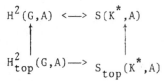

E becomes in this way a topological group, after which it is clear that A is a closed subgroup and that π induces an isomorphism of E/A onto G. Recalling the axioms of 16.1 for neighborhoods of 1, it is enough to show that $(\hat{x},\hat{y}) \longmapsto \hat{x}^{-1}\hat{y}$ is continuous at (1,1) for all $\hat{x},\hat{y}$, and that all inner automorphisms $\mathrm{Int}\,\hat{x}$ are continuous at 1.

Lemma 19.7 implies readily that $\mathrm{Int}\,\hat{x}$ is continuous at 1 for $\hat{x} \in \hat{X}_{\pm\alpha}$, $\hat{X}_{\pm\beta}$, hence for all $\hat{x} \in E$. On the other hand, $(\hat{x},\hat{y}) \longmapsto \hat{x}^{-1}\hat{y}$ is clearly continuous at (1,1) for $\hat{x} \in \hat{U}\hat{H}$, $\hat{y} \in E$. When $\hat{x} \in \hat{X}_\gamma \subset \hat{U}^-$, we conjugate by $\hat{w} \in \hat{W}$ to get back into $\hat{U}$, then use the identity $(\hat{w}\,\hat{x}\,\hat{w}^{-1})^{-1}\,\hat{y} = \hat{w}(\hat{x}^{-1}(\hat{w}^{-1}\hat{y}\,\hat{w}))\hat{w}^{-1}$ to obtain continuity of this map when $\hat{x} \in \hat{X}_\gamma$, $\hat{y} \in E$. The product decomposition of Ω_E then implies continuity of the map at (1,1) for all $\hat{x}, \hat{y} \in E$. □

§20. Moore's theory

20.1 Topological Steinberg symbols

If K is an infinite topological field and A an abelian topological group (written multiplicatively), we have defined a topological Steinberg symbol to be a bilinear map $c: K^* \times K^* \to A$ satisfying: $c(x,1-x) = 1$ for $x \neq 1$; c is continuous; $(x,y) \longmapsto c(x,1+xy)$ is continuous at $(0,0)$, where by convention $c(1,0) = c(0,1) = 1$. The set $S_{top}(K^*,A)$ of all such symbols is an abelian group under the obvious product. From 19.8 we have a diagram for $G = SL(n,K)$ (proof given for $n = 3$):

$$\begin{array}{ccccc} H^2(G,A) & \longleftrightarrow & S(K^*,A) & \cong & \mathrm{Hom}(\pi_1(G),A) \\ \text{forget}\ \uparrow & & \uparrow\ \text{incl} & & \\ H^2_{top}(G,A) & \leftrightarrow & S_{top}(K^*,A) & & \end{array}$$

This diagram suggests the introduction of a sort of topological fundamental group, which will be done in 20.3. But first we want to describe some important examples of topological Steinberg symbols.

Let $K = \mathbb{R}$, with the usual topology, and define $(\ ,\ )_\infty$:

$\mathbb{R}^* \times \mathbb{R}^* \to \{\pm 1\}$ by the rule

$$(a,b)_\infty = \begin{cases} 1 & \text{if } a > 0 \text{ or } b > 0 \\ -1 & \text{if } a < 0 \text{ and } b < 0 \end{cases}.$$

Here $\{\pm 1\}$ has the discrete topology. Bilinearity is easy to verify. Moreover, a and $1-a$ cannot both be negative, so $(a,1-a)_\infty = 1$. As to continuity, the inverse image of either 1 or -1 is a union of (open) quadrants in $\mathbb{R}^* \times \mathbb{R}^*$. Finally, we assert that $(a,1+ab)_\infty \to 1$ as $(a,b) \to (0,0)$: when (a,b) is sufficiently close to $(0,0)$, $1+ab > 0$ and hence $(a,1+ab)_\infty = 1$. So the axioms are satisfied.

In fact, the symbol just defined is essentially the only topological Steinberg symbol on $\mathbb{R}^*$. Given a symbol $c:\mathbb{R}^* \times \mathbb{R}^* \to A$, we clearly have $c(3,-2) = 1 = c(3,-3)$. Bilinearity then forces $c(3, (-2)^i(-3)^j) = 1$ for all $i, j \in \mathbb{Z}$. But the rational numbers of the form $(-2)^i(-3)^j$ are easily seen to be a dense subgroup of $\mathbb{R}^*$. So continuity forces $c(3,b) = 1$ for all $b \in \mathbb{R}^*$. A similar argument, starting with $c(4,-3) = 1 = c(4,-4)$, shows that $c(4,b) = 1$ for all b. Since the rationals of the form $3^i 4^j$ $(i,j \in \mathbb{Z})$ form a dense subgroup of the positive reals, we get $c(a,b) = 1$ for $a > 0$, b arbitrary. Next let $a < 0$, $b < 0$, and write $b = -d^2$ for some $d > 0$, so $c(a,b) = c(a,-1)\, c(a,d)^2 = c(a,-1)$, forcing $c(a,b) = c(-1,-1)$ and $c(a,b)^2 = 1$. We then have a commutative triangle:

$$\begin{array}{ccc} \mathbb{R}^* \times \mathbb{R}^* & \xrightarrow{\ c\ } & A \\ & \searrow_{(\ ,\ )_\infty} & \uparrow \\ & & \{\pm 1\} \end{array}$$

When $K = \mathbb{C}$, it is not hard to show that there are no nontrivial topological Steinberg symbols (exercise). This becomes relevant when one deals with number fields having $\mathbb{C}$ as a completion.

Next let $K = \mathbb{Q}_p$. We shall define a symbol (when $p \neq 2$) denoted $(\ ,\ )_p$, taking values in the group μ_p of roots of unity in $\mathbb{Q}_p$, which may be identified with $(\mathbb{Z}/p\mathbb{Z})^* \cong (\mathbb{Z}_p/p\mathbb{Z}_p)^*$. Any $a \in \mathbb{Q}_p^*$

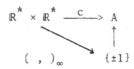

has a unique factorization $a = p^i u$, where $i \in \mathbb{Z}$ and $u \in U$ = group of units of $\mathbb{Z}_p$. Let $\delta: U \to (\mathbb{Z}_p/p\mathbb{Z}_p)^*$ be the natural homomorphism. If $a = p^i u$, $b = p^j v$, let $(a,b)_p = \delta((-1)^{ij} u^j/v^i)$. It is not hard to verify that this defines a topological Steinberg symbol (consider separately the cases $i > 0$, $i < 0$, $i = 0$ for the special continuity axiom).

This "tame" symbol $(\ ,\ )_p$ is intimately related to the familiar Legendre symbol defined by $\left(\frac{x}{p}\right) = 1$ (resp. -1) if x is a quadratic residue (resp. nonresidue) mod p. The Legendre symbol makes sense for units x of $\mathbb{Z}_p$ via the canonical map $\mathbb{Z}_p/p\mathbb{Z}_p \to \mathbb{Z}/p\mathbb{Z}$. Since p is odd, the target group of $(\ ,\ )_p$ has even order and therefore has $\{\pm 1\}$ as a quotient. The resulting composite map takes $(a,b) \in Q_p^* \times Q_p^*$ to $(-1)^{ij(p-1)/2}$ times $\left(\frac{u}{p}\right)^j \left(\frac{v}{p}\right)^i$, where $a = p^i u$, $b = p^j v$ as above. The situation differs when $p = 2$. Here we define $(a,b)_2$ to be -1 raised to the exponent $\varepsilon(u)\,\varepsilon(v) + i\,\omega(v) + j\omega(u)$, where $\varepsilon(u)$ is the class of $(u-2)/2$ mod 2 and where $\omega(u)$ is the class of $(u^2-1)/8$ mod 2. (For all of this see Serre [4, Chapter III].)

The classical quadratic reciprocity law can now be expressed as follows:

$$\prod_{p \leq \infty} (a,b)_p^{m_p/m} = 1, \quad \text{for} \quad a, b \in Q.$$

Here m_p is the order of μ_p (when we set $\mathbb{R} = Q_\infty$, $m_\infty = 2$), and m is the number of roots of unity in Q (namely, 2).

20.2 Local and global theorems

Moore showed that the symbols for $\mathbb{R}$ and Q_p described in 20.1, along with the reciprocity law relating all of them, are in a certain sense unique. (More generally, he dealt with arbitrary local and global fields, where there are analogous "norm residue symbols" and reciprocity laws.)

LOCAL THEOREM. *Let* K *be either* $\mathbb{R} = \mathbb{Q}_\infty$ *or* $\mathbb{Q}_p$, *p prime. If* $c: K^* \times K^* \to A$ *is a topological Steinberg symbol, there is a unique dotted map making the triangle commute:*

$$\begin{array}{ccc} K^* \times K^* & \xrightarrow{\;c\;} & A \\ & \searrow^{(\;,\;)_p} & \uparrow \vdots \\ & & \mu_p \end{array}$$

Thus $S_{top}(K,A) \cong \mathrm{Hom}(\mu_p, A)$.

For the proof, consult Moore [1, §3 - §6] or Milnor [1,Appendix].

GLOBAL THEOREM. *The following sequence is exact:*

$$\mathbb{Q}^* \otimes \mathbb{Q}^* \xrightarrow{\;\phi\;} \bigoplus_{p\leq\infty} \mu_p \xrightarrow{\;\psi\;} \mu_{\mathbb{Q}} \dashrightarrow 1. \qquad \mu_{\mathbb{Q}} = \{\pm 1\}$$

Here $\phi(a \otimes b)$ *has* $(a,b)_p$ *as* p^{th} *component, while* $\psi(\ldots\zeta_p\ldots) = \prod \zeta_p^{\,m_p/m}$.

In this theorem, $\psi\phi = 1$ just expresses the classical quadratic reciprocity law for $\mathbb{Q}$, while $\mathrm{Ker}\,\psi \subset \mathrm{Im}\,\phi$ expresses the essential uniqueness of this law. That ϕ is well defined follows from the fact that $a, b \in \mathbb{Q}^*$ are p-adic units for almost all p, and then $(a,b)_p = 1$. Moore's original proof [1, §7] relied somewhat on results of Bass-Milnor-Serre[1], but subsequently Chase and Waterhouse [1] found a more selfcontained proof (which still uses such tools as Dirichlet's theorem on primes in arithmetic progressions).

20.3 Central extensions of locally compact groups

Moore's study of the cohomology of locally compact groups led him to introduce notions of "covering" and "fundamental group" in the context of (topological) central extensions. For technical reasons, the groups are always assumed to be separable, i.e., to have a countable basis of neighborhoods of 1; this assumption is automatically satisfied by the groups of interest to us.

In the following brief outline of Moore's theory, the groups

G,A,E,... are locally compact, with A abelian, and "central extension" means "topological central extension".

Call G _simply connected_ if for any central extension $1 \to A \to E \to G \to 1$, there is a unique continuous homomorphism $G \to E$ splitting the extension. A central extension $1 \to A \to E \to G \to 1$ is called a _covering_ of G provided (G,G) is dense in G and (E,E) is dense in E. (This is a kind of connectedness assumption.) A covering is called _universal_ if the group E is simply connected.

THEOREM. (a) _Given G, there exists (up to isomorphism) at most one universal covering of G. (The kernel is then denoted_ $\pi_1^{top}(G)$ _and called the fundamental group of G.)_

(b) _Let_ $1 \to \pi_1^{top}(G) \to E_0 \to G \to 1$ _be a universal covering of G. Given any central extension_ $1 \to A \to E \to G \to 1$, _there is a unique (continuous) homomorphism_ $E_0 \to E$ _making the obvious diagram commute._

Universal coverings exist in many cases of interest, as we shall see below. When G is discrete and G = (G,G), there always exists a universal covering of G. For G = SL(n,K), we constructed such a covering explicitly in §18 and called it St(n,K); in that case $\pi_1^{top}(G)$ is the same as our earlier $\pi_1(G)$.

20.4 The fundamental group in the local case

THEOREM. _Let K be_ $\mathbb{R} = \mathbb{Q}_\infty$ _or_ $\mathbb{Q}_p$, $G = SL(n,K)$, $n \geq 3$. _Then G has a universal covering, with_ $\pi_1^{top}(G) \cong \mu_K$, _the roots of unity in K._

This is a fairly direct consequence of the Local Theorem 20.2, which shows that $S_{top}(K^*,A) \cong Hom(\mu_K,A)$, together with the explicit construction of topological central extensions in §18-19. The point is that the Steinberg symbol corresponding to the universal covering of G is just the norm residue symbol described in 20.1.

There is obvious analogue of the above theorem for other simple, simply connected algebraic groups, and for completions of number fields other than $\mathbb{Q}$. When $\mathbb{C}$ occurs as a completion, we get $\pi_1^{top}(G) = 1$: G is "simply connected" in Moore's sense as well as in the sense of algebraic (or complex Lie) groups. There is one exceptional case: when G is a symplectic group over $\mathbb{R}$ (including $SL(2,\mathbb{R})$), $\pi_1^{top}(G)$ is infinite cyclic. This reflects the usual behavior of the fundamental groups of real Lie groups, known since E. Cartan's work.

Consider again $G = SL(n,\mathbb{Q}_p)$, $n \geq 3$. Let H be the compact open subgroup $SL(n,\mathbb{Z}_p)$. When the residue field $\mathbb{Z}/p\mathbb{Z}$ is big enough (in our case when $p \geq 4$), it can be shown that $H = (H,H)$. Moreover, H then possesses a universal covering D, with $\pi_1^{top}(H)$ being a projective limit of finite p-groups. The universal property of D provides a diagram:

$$\begin{array}{ccccccccc} 1 & \to & \pi_1^{top}(H) & \to & D & \to & H & \to & 1 \\ & & \downarrow & & \downarrow & & \downarrow & & \\ 1 & \to & A & \to & E' & \to & H & \to & 1 \\ & & \downarrow & & \downarrow & & \downarrow & & \\ 1 & \to & \mu_p & \to & E & \to & G & \to & 1 \end{array}$$

Here E is the universal covering of G, E' the inverse image of H. Since μ_p has order relatively prime to p for most p (in our case for $p > 2$), whereas $\pi_1^{top}(H)$ is a pro-p-group, the middle extension in the diagram splits for most p.

20.5 Restricted products

Consider the restricted product G of a family $\{G_i\}$ of locally compact groups relative to compact open subgroups H_i (defined for almost all i). G is itself locally compact, so we can ask about its fundamental group, if one exists.

THEOREM. *With G, G_i, H_i as above, suppose each G_i has a universal covering E_i, and suppose almost every H_i has a universal covering L_i. Then G has a universal covering, namely the restricted product of the E_i relative to the canonical images of the L_i in*

E_i. Thus $\pi_1^{top}(G)$ is the restricted product of the fundamental groups of the G_i relative to the canonical images of the fundamental groups of the H_i.

We can apply this result of Moore to the family $G_p = SL(n,Q_p)$, $p \leq \infty$, relative to the subgroups $H_p = SL(n,\mathbb{Z}_p)$, to obtain a universal covering for the restricted product $G_{\mathbb{A}} = SL(n,\mathbb{A})$, where $\mathbb{A}$ is the ring of adeles. If we omit the prime ∞ from the picture, we have a similar situation for $\bar{G} = G_{\mathbb{A}^f}$, where $\mathbb{A}^f$ is the ring of finite adeles. From the local results 20.4 we conclude:

$$\pi_1^{top}(G_{\mathbb{A}}) \cong \bigoplus_{p\leq\infty} \mu_p$$

$$\pi_1^{top}(G_{\mathbb{A}^f}) \cong \bigoplus_{p\neq\infty} \mu_p$$

20.6 Relative coverings

Finally, we have to take into account the possible splitting of a central extension of $G_{\mathbb{A}}$ or $G_{\mathbb{A}^f}$ on restriction to the discrete (resp. dense), diagonally embedded subgroup $G_Q = SL(n,Q)$.

For the moment, let G be just a locally compact group, with (G,G) dense in G, and let $j:H \to G$ be a continuous injection (not necessarily a homeomorphism). Associated with each central extension E of G is a central extension $j^*(E)$ of H defined in the obvious way; we say the extension is trivial with respect to H if $j^*(E)$ splits. A covering E of G is a relative universal covering with respect to H if it is trivial with respect to H and maps uniquely into each such central extension of G. Up to isomorphism, Moore shows that at most one such relative universal covering exists; then the kernel $\pi_1(G,H)$ is called the relative fundamental group.

THEOREM. With $j:H \to G$ as above, suppose both G and H possess universal coverings. Then there exists a relative universal covering, with $\pi_1(G,H)$ the quotient of $\pi_1^{top}(G)$ by the closure of the canonical image of $\pi_1^{top}(H)$. Moreover, $\pi_1(G,H)$ is discrete if H is a compact open subgroup of G.

In our situation, j maps $G_{\mathbb{Q}}$ diagonally into either G_A or G_{A^f}. Matsumoto's theorem implies that $\pi_1(G_{\mathbb{Q}})$ is isomorphic to the quotient of $\mathbb{Q}^* \otimes \mathbb{Q}^*$ by the subgroup generated by all $a \otimes (1-a)$, cf. exercise in 19.2. The map ϕ in Moore's Global Theorem 20.2 sends all $a \otimes (1-a)$ to 1, hence induces the exact sequence:

$$\begin{array}{ccccc} \pi_1(G_{\mathbb{Q}}) & \xrightarrow{\phi} & \bigoplus_{p\leq\infty} \mu_p & \xrightarrow{\psi} & \mu_{\mathbb{Q}} = \{\pm 1\} \longrightarrow 1 \\ & \searrow^{\phi'} & \downarrow{\scriptstyle proj} & & \\ & & \bigoplus_{p\neq\infty} \mu_p & & \end{array}$$

We can also define ϕ' to make a commutative triangle. Observe that the maps ϕ, ϕ' are none other than the canonical maps:

$$\begin{array}{ccc} \pi_1(G_{\mathbb{Q}}) & \longrightarrow & \pi_1^{top}(G_A) \\ & \searrow & \\ & & \pi_1^{top}(G_{A^f}) \end{array}$$

From this we conclude immediately, since $\mu_\infty = \{\pm 1\}$ is the kernel of the projection: $\pi_1(G_A, G_{\mathbb{Q}}) = \mathbb{Z}/2\mathbb{Z}$, $\pi_1(G_{A^f}, G_{\mathbb{Q}}) = 1$.

Remarks. (1) Actually $\operatorname{Ker} \phi = 1$ above (see Milnor [1, §11]). For an arbitrary number field, $\operatorname{Ker} \phi$ is at any rate finite (see Garland [1]).

(2) If we were working over a totally imaginary number field K, we would not have any occurrence of μ_∞ in the above discussion, so we would arrive instead at $\pi_1(G_{A^f}, G_K) = \mu_K$. (This leads to a negative answer for the Congruence Subgroup Problem.)

(3) If we worked with a symplectic group, the fact that the fundamental group in the real case is infinite would not in any way affect

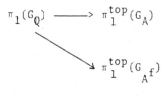

the determination of the relative fundamental group of G_{A^f}. So the Congruence Subgroup Problem does not have an essentially different solution in this case (barring $SL(2,K)$).

20.7 The congruence kernel revisited

It remains only to compare the conclusion of 20.6 with our earlier formulation 17.6 of the Congruence Subgroup Problem in terms of the exact sequence (*) $1 \to C \to \hat{G} \overset{\pi}{\to} \overline{G} \to 1$. Here $\overline{G} = G_{A^f}$. We saw that (*) is a central extension, trivial with respect to $G_{\mathbb{Q}}$, and that it is universal relative to these properties. So it must agree with the relative universal covering discussed in 20.6. In particular, $C \cong \pi_1(G_{A^f}, G_{\mathbb{Q}}) = 1$. So for $SL(n,\mathbb{Q})$, $n \geq 3$, the Congruence Subgroup Problem has an affirmative solution! (The same would be true for other simple, simply connected algebraic groups of rank at least 2, and other number fields, except for totally imaginary fields where instead $C \cong \mu_K$.)

SUGGESTIONS FOR FURTHER READING

By its nature, the subject of arithmetic groups reaches out in many directions and has ill-defined boundaries. So it would be almost impossible to assemble a comprehensive bibliography. Instead, I have listed below some works which either develop or complement the topics treated in these notes. No claims of completeness (or even balance) are intended. Most of the papers cited presuppose some sophistication in the use of Lie groups or linear algebraic groups; but in his monograph, Borel [5] does attempt to ease the transition from special cases to general theory. Here are a few topics which the reader may wish to explore further:

Reduction theory for arithmetic subgroups of semisimple Lie groups is treated in Borel [5]; cf. Harder [1] for the function field case. Earlier work along these lines is also well worth consulting, e.g., Borel, Harish-Chandra [1], Godement [1], Mostow, Tamagawa [1]. The measure of a fundamental set is studied in various articles in Borel, Mostow [1], notably Langlands [1], as well as in Weil [1], Mars [1] - [3].

Cohomology of arithmetic groups can be looked at from many points of view. The work of Borel leads in a natural way to the results of Borel, Serre [1], as was mentioned briefly in 13.5. There are many other papers on cohomology, of which the following list is just a sample: Borel [6], [7], Borel, Wallach [1], Garland [2], Harder [2], Kazhdan [1], Raghunathan [1], Schwermer [1], Soulé [1].

As indicated in 13.4, reduction theory yields information about finite generation or finite presentation of arithmetic groups. Relevant papers include: Borel [1], Kneser [1], Behr [1] - [3], Stuhler [1]. Steinberg's work on central extensions (cf. §18 above) leads in a different way to explicit presentations of Chevalley groups over fields as well as over rings like $\mathbb{Z}$: see Steinberg [1], [2], Deodhar [1], Milnor [1], Behr [5].

The modular group $SL(2,\mathbb{Z})$ or $PSL(2,\mathbb{Z})$ has a life of its own. From the extensive literature we cite just a few sources, selected at random: Reiner [1], Newman [1], Jones [1]. Of special interest are congruence (and non-congruence) subgroups in SL_2 over the integers of various global fields, cf. Serre [3], [5], Mel'nikov [1].

The congruence subgroup problem has been attacked in greater and greater generality, but is still not entirely resolved for groups of rank at least 2 over global fields. For $SL(n,\mathbb{Z})$, $n \geq 3$, independent solutions were found by Bass, Lazard, Serre [1] and Mennicke [1].

Then Bass, Milnor, Serre [1] (cf. the exposition in Serre [2]) treated special linear and symplectic groups over arbitrary number fields, the case of SL_2 being handled separately by Serre [3]. Using results of Moore [1], Matsumoto [1] finished off the split (Chevalley type) groups. Non-split groups have been studied by a number of authors: Vaserstein [1] - [3], Kneser [5], Deodhar [1], Raghunathan [4]. The connections with algebraic K-theory have also been thoroughly explored, cf. Milnor [1], Keune [1].

In an entirely different direction, it is natural to ask whether every lattice in a semisimple Lie group G (discrete subgroup H for which G/H has finite invariant measure) is defined arithmetically, relative to some rational structure on G. In rank 1 there are known to be exceptions (cf. Vinberg [1], Mostow [3]), but in rank ≥ 2 (suitably formulated) it turns out that all lattices are indeed arithmetic. Partial results in this direction were obtained by Prasad, Raghunathan [1], Raghunathan [5] (see Raghunathan [3] for further background on these matters). But the most general results are due to Margulis [1] - [3]; Tits [3] provides a very helpful exposition. For related questions about rigidity of lattices, see Mostow [1], [2], Prasad [1], [2].

REFERENCES

H. Bass, M. Lazard, J.-P. Serre

1. Sous-groupes d'indices finis dans SL(n,Z), Bull. Amer. Math. Soc. 70 (1964), 385-392

H. Bass, J. Milnor, J.-P. Serre

1. Solution of the congruence subgroup problem for SL_n ($n \geq 3$) and Sp_{2n} ($n \geq 2$), Inst. Hautes Études Sci. Publ. Math. 33 (1967), 59-137

H. Behr

1. Über die endliche Definierbarkeit von Gruppen, J. Reine Angew. Math. 211 (1962), 116-122
2. Über die endliche Definierbarkeit verallgemeinerter Einheitengruppen, II, Invent. Math. 4 (1967), 265-274
3. Endliche Erzeugbarkeit arithmetischer Gruppen über Funktionenkörpern, Invent. Math. 7 (1969), 1-32
4. Zur starken Approximation in algebraischen Gruppen über globalen Körpern, J. Reine Angew. Math. 229 (1968), 107-116
5. Explizite Präsentation von Chevalleygruppen über Z, Math. Z. 141 (1975), 235-241

A. Borel

1. Arithmetic properties of linear algebraic groups, pp. 10-22, Proc. Intl. Congr. Math., Stockholm, 1962
2. Some finiteness properties of adele groups over number fields, Inst. Hautes Études Sci. Publ. Math. 16 (1963), 5-30
3. Density and maximality of arithmetic subgroups, J. Reine Angew. Math. 224 (1966), 78-89
4. Linear Algebraic Groups, Notes by H. Bass, W.A. Benjamin, New York, 1969
5. Introduction aux groupes arithmétiques, Hermann, Paris, 1969
6. Stable real cohomology of arithmetic groups, Ann. Sci. École Norm. Sup. 7 (1974), 235-272
7. Cohomologie de sous-groupes discrets et représentations de groupes semi-simples, Soc. Math. France, Astérisque 32-33 (1976), 73-112

A. Borel, G. Harder

1. Existence of discrete cocompact subgroups of reductive groups over local fields, J. Reine Angew. Math. 298 (1978), 53-64

A. Borel, Harish-Chandra

1. Arithmetic subgroups of algebraic groups, Ann. of Math. 75 (1962), 485-535

A. Borel, G.D. Mostow, ed.

1. Algebraic Groups and Discontinuous Subgroups, Proc. Symp. Pure Math. 9, Amer. Math. Soc., Providence, RI, 1966

A. Borel, J.-P. Serre

1. Corners and arithmetic groups, Comment. Math. Helv. 48 (1973), 436-491

A. Borel, N. Wallach

1. Continuous cohomology, discrete subgroups and representations of reductive groups, Ann. of Math. Studies No. 94, Princeton Univ. Press, 1980

N. Bourbaki

1. General topology, Addison-Wesley, Reading, MA, 1966
2. Groupes et algèbres de Lie, Chapters 4-6, Hermann, Paris, 1968

N. Bourbaki
3. Intégration, Chapters 7-8, Hermann, Paris, 1963

F. Bruhat, J. Tits
1. Groupes algébriques simples sur un corps local, pp. 23-36, Proc. Conf. on Local Fields, Springer, Berlin, 1967
2. Groupes réductifs sur un corps local, I, Inst. Hautes Études Sci. Publ. Math. 41 (1972), 5-252

R.W. Carter
1. Simple Groups of Lie Type, Wiley-Interscience, London, 1972

J.W.S. Cassels
1. Global fields, pp. 42-84 in Cassels, Fröhlich [1]

J.W.S. Cassels, A. Fröhlich, ed.
1. Algebraic Number Theory, Thompson Book Co., Washington, D.C., 1967

S.U. Chase, W.C. Waterhouse
1. Moore's theorem on uniqueness of reciprocity laws, Invent. Math. 16 (1972), 267-270

P. Deligne
1. Extensions centrales non résiduellement finies de groupes arithmétiques, C.R. Acad. Sci. Paris Sér. A-B 287 (1978), A203 -A208

V.V. Deodhar
1. On central extensions of rational points of algebraic groups, Amer. J. Math. 100 (1978), 303-386

H. Garland
1. A finiteness theorem for K_2 of a number field, Ann. of Math. 94 (1971), 534-548
2. p-adic curvature and the cohomology of discrete subgroups, Ann. of Math. 97 (1973), 375-423

S. Gelbart
1. Automorphic Forms on Adele Groups, Ann. of Math. Studies No. 83, Princeton Univ. Press, 1975

I.M. Gel'fand, M.I. Graev, I.I. Pyatetskii-Shapiro
1. Representation Theory and Automorphic Functions, W.B. Saunders, Philadelphia, 1969

R. Godement
1. Domaines fondamentaux des groupes arithmétiques, Sém. Bourbaki 1962/63, Exp. 257

L.J. Goldstein
1. Analytic Number Theory, Prentice-Hall, Englewood Cliffs, N.J., 1971

P.R. Halmos
1. Measure Theory, Van Nostrand, Princeton, 1950

G. Harder
1. Minkowskische Reduktionstheorie über Funktionenkörpern, Invent. Math. 7 (1969), 33-54
2. A Gauss-Bonnet formula for discrete arithmetically defined groups, Ann. Sci. École Norm. Sup. 4 (1971), 409-455

P.J. Higgins
1. An Introduction to Topological Groups, London Math. Soc. Lect. Note Series 15, Cambridge Univ. Press, 1974

H. Hijikata
1. On the structure of semi-simple algebraic groups over valuation fields, I, Japan J. Math. 1 (1975), 225-300

J.E. Humphreys
1. Linear Algebraic Groups, Grad. Texts in Math. 21, Springer, Berlin-Heidelberg-New York, 1975

N. Iwahori, H. Matsumoto
1. On some Bruhat decomposition and the structure of the Hecke rings of p-adic Chevalley groups, Inst. Hautes Études Sci. Publ. Math. 25 (1965), 5-48

Y. Ihara
1. On discrete subgroups of the two by two projective linear group over p-adic fields, J. Math. Soc. Japan 18 (1966), 219-235

G.A. Jones
1. Triangular maps and non-congruence subgroups of the modular group, Bull. London Math. Soc. 11 (1979), 117-123

D.A. Kazhdan
1. Connection of the dual space of a group with the structure of its closed subgroups, Functional Anal. Appl. 1 (1967), 63-65

F. Keune
1. (t^2-t)-reciprocities on the affine line and Matsumoto's theorem, Invent. Math. 28 (1975), 185-192

M. Kneser
1. Erzeugende und Relationen verallgemeinerter Einheitengruppen, J. Reine Angew. Math. 214/215 (1964), 345-349
2. Starke Approximation in algebraischen Gruppen I, J. Reine Angew. Math. 218 (1965), 190-203
3. Strong approximation, pp. 187-196 in Borel, Mostow [1]
4. Semi-simple algebraic groups, pp. 251-265 in Cassels, Fröhlich [1]
5. Normal subgroups of integral orthogonal groups, pp. 67-71, Algebraic K-Theory and its Geometric Applications, Lect. Notes in Math. 108, Springer, Berlin, 1969

S. Lang
1. Algebraic Number Theory, Addison-Wesley, Reading, MA, 1970

R.P. Langlands
1. The volume of the fundamental domain for some arithmetical subgroups of Chevalley groups, pp. 143-148 in Borel, Mostow [1]

I.G. Macdonald
1. Spherical functions on a group of p-adic type, Publ. of Ramanujan Institute No. 2, Univ. Madras, 1971

G.A. Margulis
1. Non-uniform lattices in semisimple algebraic groups, pp. 371-553, Lie Groups and their Representations, ed. I.M. Gel'fand, Halsted, New York, 1975
2. Arithmetic properties of discrete subgroups, Russian Math. Surveys 29 (1974), 107-156
3. Discrete groups of motions of manifolds of nonpositive curvature, Amer. Math. Soc. Transl. (Ser. 2) 109 (1977), 33-45 [Russian original appears in proceedings of 1974 Intl. Congr. Math., Vancouver]
4. Cobounded subgroups of algebraic groups over local fields, Functional Anal. Appl. 11 (1977), 119-128

J.G.M. Mars

1. Les nombres de Tamagawa de certains groupes exceptionnels, Bull. Soc. Math. France 94 (1966), 97-140
2. Solutions d'un problème posé par A. Weil, C.R. Acad. Sci. Paris Sér. A-B 266 (1968), A484-A486
3. The Tamagawa number of ${}^2A_n$, Ann. of Math. 89 (1969), 557-574

H. Matsumoto

1. Sur les sous-groupes arithmétiques des groupes semi-simples déployés, Ann. Sci. École Norm. Sup. 2 (1969), 1-62

O.V. Mel'nikov

1. Congruence kernel of the group $SL_2(\mathbb{Z})$, Soviet Math. Dokl. 17 (1976), 867-870

J. Mennicke

1. Finite factor groups of the unimodular group, Ann. of Math. 81 (1965), 31-37
2. On Ihara's modular group, Invent. Math. 4 (1967), 202-228

J. Milnor

1. Introduction to Algebraic K-Theory, Ann. of Math. Studies No. 72, Princeton Univ. Press, 1971

C.C. Moore

1. Group extensions of p-adic and adelic linear groups, Inst. Hautes Études Sci. Publ. Math. 35 (1969), 5-70

G.D. Mostow

1. Strong rigidity of locally symmetric spaces, Ann. of Math. Studies No. 78, Princeton Univ. Press, 1973
2. Discrete subgroups of Lie groups, Advances in Math. 16 (1975), 112-123
3. Existence of a nonarithmetic lattice in SU(2,1), Proc. Nat. Acad. Sci. U.S.A. 75 (1978), 3029-3033

G.D. Mostow, T. Tamagawa

1. On the compactness of arithmetically defined homogeneous spaces, Ann. of Math. 76 (1961), 446-463

M. Newman

1. Maximal normal subgroups of the modular group, Proc. Amer. Math. Soc. 19 (1968), 1138-1144

O.T. O'Meara

1. Introduction to Quadratic Forms, Springer, Berlin, 1963

W. Page

1. Topological Uniform Structures, Wiley, New York, 1978

V.P. Platonov

1. Adele groups and integral representations, Math. USSR-Izv. 3 (1969), 147-154
2. The problem of strong approximation and the Kneser-Tits conjecture for algebraic groups, Math. USSR-Izv. 3 (1969), 1139-1147; addendum, ibid. 4 (1970), 784-786
3. On the maximality problem for arithmetic groups, Soviet Math. Dokl. 12 (1971), 1431-1435
4. On the genus problem in arithmetic groups, Soviet Math. Dokl. 12 (1971), 1503-1507
5. The arithmetic theory of linear algebraic groups and number theory, Proc. Steklov Inst. Math. 132 (1973), 184-191
6. Arithmetical and structural problems in linear algebraic groups, Amer. Math. Soc. Transl. (Ser. 2) 109 (1977), 21-26 [Russian original in Proc. Intl. Congr. Math., Vancouver 1974]

V.P. Platonov, A.A. Bondarenko, A.S. Rapinčuk
1. Class number and class group of algebraic groups, Math. USSR-Izv. 13 (1979)

V.P. Platonov, M.V. Milovanov
1. Determination of algebraic groups by arithmetic subgroups, Soviet Math. Dokl. 14 (1973), 331-335

G. Prasad
1. Strong rigidity of $\mathbb{Q}$-rank 1 lattices, Invent. Math. 21 (1973), 255-286
2. Discrete subgroups isomorphic to lattices in semisimple Lie groups, Amer. J. Math. 98 (1976), 241-261
3. Strong approximation for semi-simple groups over function fields, Ann. of Math. 105 (1977), 553-572
4. Lattices in semisimple groups over local fields, pp. 285-356, Studies in Algebra and Number Theory, Academic Press, New York, 1979

G. Prasad, M.S. Raghunathan
1. Cartan subgroups and lattices in semi-simple groups, Ann. of Math. 96 (1972), 296-317

M.S. Raghunathan
1. Cohomology of arithmetic subgroups of algebraic groups, I, Ann. of Math. 86 (1967), 409-424; II, ibid. 87 (1968), 279-304
2. A note on quotients of real algebraic groups by arithmetic subgroups, Invent. Math. 4 (1968), 318-335
3. Discrete Subgroups of Lie Groups, Springer, Berlin, 1972
4. On the congruence subgroup problem, Inst. Hautes Études Sci. Publ. Math. 46 (1976), 107-161
5. Discrete groups and $\mathbb{Q}$-structures on semi-simple Lie groups, pp. 225-321, Discrete subgroups of Lie groups and applications to moduli (Internat. Colloq., Bombay, 1973), Oxford Univ. Press, Bombay, 1975

I. Reiner
1. Normal subgroups of the unimodular group, Illinois J. Math. 2 (1958), 142-144

F.A. Richen
1. Modular representations of split BN pairs, Trans. Amer. Math. Soc. 140 (1969), 435-460

A. Robert
1. Des adèles: pourquoi, Enseignement Math. 20 (1974), 133-145

J. Rohlfs
1. Über maximale arithmetisch definierte Gruppen, Math. Ann. 234 (1978), 239-252

J. Schwermer
1. Sur la cohomologie des sous-groupes de congruence de $SL_3(\mathbb{Z})$, C.R. Acad. Sci. Paris Sér. A-B 283 (1976), A817-A820

J.-P. Serre
1. Lie Algebras and Lie Groups, W.A. Benjamin, New York, 1965
2. Groupes de congruence, Sém. Bourbaki 1966/67, Exp. 330
3. Le problème des groupes de congruence pour SL_2, Ann. of Math. 92 (1970), 489-527
4. A Course in Arithmetic, Grad. Texts in Math. 7, Springer, Berlin-Heidelberg-New York, 1973
5. Arbres, amalgames, SL_2, Soc. Math. France, Astérisque 46 (1977)

C. Soulé

1. The cohomology of $SL_3(\mathbb{Z})$, Topology 17 (1978), 1-22

R. Steinberg

1. Générateurs, relations et revêtements de groupes algébriques, pp. 113-127, Colloq. Théorie des Groupes Algébriques (Bruxelles, 1962), Gauthier-Villars, Paris, 1962
2. Lectures on Chevalley groups, Yale Univ. Math. Dept., 1968

U. Stuhler

1. Zur Frage der endlichen Präsentierbarkeit gewisser arithmetischer Gruppen im Funktionenkörperfall, Math. Ann. 224 (1976), 217-232

T. Tamagawa

1. On discrete subgroups of p-adic algebraic groups, pp. 11-17, Arithmetical Algebraic Geometry, ed. O.F.G. Schilling, Harper & Row, New York, 1965

J.T. Tate

1. Fourier analysis in number fields and Hecke's zeta-functions, pp. 305-347 in Cassels, Fröhlich [1]

J. Tits

1. Théorème de Bruhat et sous-groupes paraboliques, C.R. Acad. Sci. Paris Sér. A-B 254 (1962), A2910-A2912
2. Systèmes générateurs de groupes de congruence, C.R. Acad. Sci. Paris Sér. A-B 283 (1976), A693-A695
3. Travaux de Margulis sur les sous-groupes discrets de groupes de Lie, Sém. Bourbaki 1975/76, Exp. 482, Lect. Notes in Math. 567, Springer, Berlin, 1977
4. Reductive groups over local fields, pp. 29-69, Proc. Symp. Pure Math. 33, Part 1, Amer. Math. Soc., Providence RI, 1979

L.N. Vaseršteĭn

1. Subgroups of finite index of a spinor group of rank $\geqslant$ 2, Math. USSR-Sb. 4 (1968), 161-166
2. The congruence problem for a unitary group of rank $\geqslant$ 2, Math. USSR-Sb. 5 (1968), 351-356
3. The structure of classical arithmetic groups of rank greater than one, Math. USSR-Sb. 20 (1973), 465-492

E.B. Vinberg

1. Discrete groups generated by reflections in Lobacevskii spaces, Math. USSR-Sb. 1 (1967), 429-444

A. Weil

1. Adeles and algebraic groups, Inst. for Advanced Study, Princeton, 1961
2. Basic Number Theory, 3rd ed., Springer, New York-Heidelberg-Berlin, 1974

INDEX